Power Maths

Year 6 Textbook 6B

Series Editor: Tony Staneff

D1242971

Ash

Ash is curious and inquisitive.

He loves to explore new concepts.

flexible

brave

determined

helpful

Flo

Astrid

Dexter

Sparks

Pearson

Contents

This tells you which page you need.

Are you ready for some more maths?

How to use this book

These pages make sure we're ready for the unit ahead. Find out what we'll be learning and brush up on your skills!

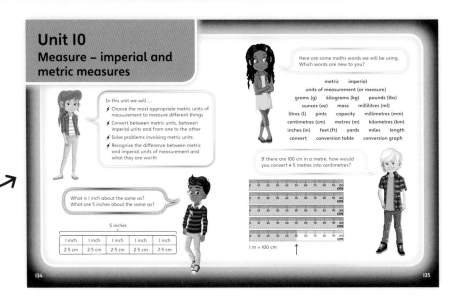

Discover

Lessons start with **Discover**.

Here, we explore new maths problems.

Can you work out how to find the answer?

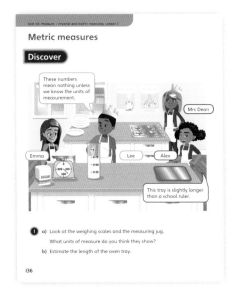

Don't be afraid to make mistakes. Learn from them and try again!

Share

Next, we share our ideas with the class.

Did we all solve the problems the same way? What ideas can you try?

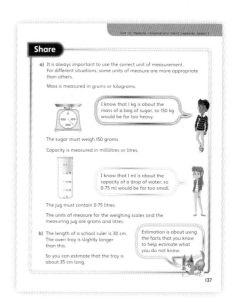

Think together

Then we have a go at some more problems together. Use what you have just learnt to help you.

We'll try a challenge too!

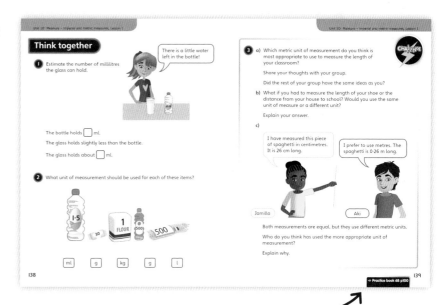

This tells you which page to go to in your **Practice Book**.

At the end of each unit there's an **End of unit check**. This is our chance to show how much we have learnt.

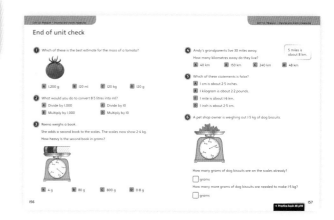

Unit 7
Decimals

In this unit we will ...

⚡ Recognise the value of each digit in a decimal number

⚡ Multiply and divide decimals by 10, 100 and 1,000

⚡ Convert between fractions and decimals

⚡ Multiply and divide decimals by single digit numbers

Do you remember using place value grids?

H	T	O	•	Tth	Hth	Thth
			•			

We will need some maths words. Have you used any of these before? What can you remember about fractions?

multiply   divide   decimal

decimal place (dp)   recurring decimal

placeholder   place value

tenths   hundredths   thousandths

products   fraction

Can you identify the value of each digit? Explain how you know to your partner.

H	T	O	•	Tth	Hth	Thth
3	0	4	•	9	0	8

Multiplying by 10, 100 and 1,000

Discover

1 **a)** Each plate has a mass of 0·3 kg. What is the mass of the 10 plates altogether?

b) Each glass has a mass of 0·15 kg. What is the total mass of all the glasses?

Share

a) There are 10 plates, so find 0·3 × 10.

T	O	•	Tth
		•	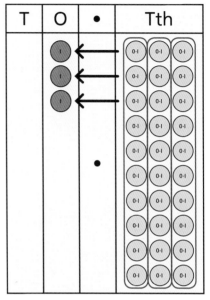

Represent 0·3.

> Exchanging the tenths means the digit moves one place to the left on the place value grid.

T	O	•	Tth
		•	

Multiply by 10.

T	O	•	Tth
		•	

Exchange each group of ten tenths.

T	O	•	Tth
		•	3

T	O	•	Tth
	3	•	3

T	O	•	Tth
	3	•	

0·3 × 10 = 3

300 g × 10 = 3,000 g

3,000 g is equivalent to 3 kg.

The mass of the 10 plates altogether is 3 kg.

> I solved it by converting to grams. I know that 0·3 kg is equivalent to 300 g.

b) There are 10 glasses, so find 0·15 × 10.

T	O	•	Tth	Hth
		•	0·1	0·01 0·01 0·01 0·01 0·01

T	O	•	Tth	Hth
		•	1	5

T	O	•	Tth	Hth
	1	•	0·1 0·1 0·1 0·1 0·1	

T	O	•	Tth	Hth
	1	•	5	

0·15 × 10 = 1·5 kg

The total mass of all the glasses is 1·5 kg.

Think together

1 A pack of pencils costs £2·30. Mr Lopez buys 100 packs.

How much does this cost in total?

H	T	O	•	Tth	Hth
			•	0·1 0·1 0·1	

$$\boxed{2} \quad \boxed{\cdot} \quad \boxed{3} \quad \boxed{0}$$

100 packs of pencils cost ☐ in total.

2 Complete these calculations.

a) 5·2 × 10 = ☐

b) 5·2 × 100 = ☐

c) 5·2 × 1,000 = ☐

d) 0·12 × 10 = ☐

e) 1·02 × 100 = ☐

f) 10·02 × 1,000 = ☐

g) 50·2 × 10 = ☐

h) 5·02 × 100 = ☐

i) 0·502 × 1,000 = ☐

3 An artist uses these tiles to make a mosaic border along a wall. The height of the mosaic is 20 tiles and the width is 500 tiles.

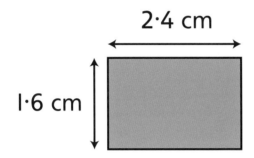

2·4 cm

1·6 cm

Each tile is 2·4 cm wide and 1·6 cm high.

What is the height and width of the whole mosaic in centimetres?

If I know 10 times a number, I can work out 20 times a number.

If I know 100 times a number, I can work out 500 times a number.

The height of the mosaic is ☐ and the width is ☐ .

→ Practice book 6B p6

Dividing by multiples of 10, 100 and 1,000

Discover

Aki

We need to make 12 m of paper chains. I need 10 of you to do this.

Mrs Dean

1 **a)** 10 children make 12 m of paper chains. They each make an equal length chain. What length do they each make?

b) There are 20 children in the class altogether. What length of paper chain would each child make if they each made an equal share of the 12 m?

Share

a) 12 ÷ 10 = ?

Find out what number is multiplied by 10 to make 12.

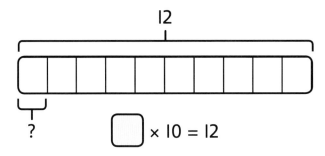

H	T	O	•	Tth	Hth
	1	2	•		

$\boxed{}$ × 10 = 12

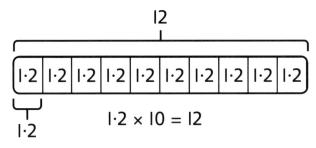

H	T	O	•	Tth	Hth
		1	•	2	

1·2 × 10 = 12

1·2 × 10 = 12   so   12 ÷ 10 = 1·2

Each child makes 1·2 m of paper chain.

> Dividing by 10 is the inverse of multiplying by 10.

b) There are twice as many parts, so divide each into two equal parts.

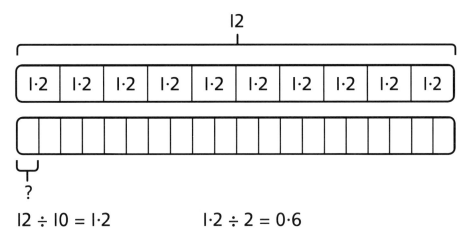

12 ÷ 10 = 1·2   1·2 ÷ 2 = 0·6

Each child would make 0·6 m of paper chain.

Think together

1 Aki shares 15 litres of juice between 100 cups.

How much juice will he pour into each cup?

H	T	O	•	Tth	Hth
	1	5	•		

$15 \div 100 = \boxed{}$

Aki will pour $\boxed{}$ litres of juice into each cup.

> I wonder if I will get the same result if I convert into ml first.

2 Reena draws this diagram to explain dividing 0·2 by 10.

O	•	Tth	Hth	Thth
	•			

Exchange each 0·1 for ten 0·01s.

→

O	•	Tth	Hth	Thth
	•			

Divide 20 counters by 10.

She says, '0·2 ÷ 10 = 0·02, and that is why the digits move to the right.'

Do you agree with Reena? Discuss her explanation with your partner.

Make or draw a similar diagram to show 12·3 ÷ 100.

3 **a)** Complete these two divisions. Which method do you think is most efficient?

$40 \div 50 = \boxed{}$

$40 \longrightarrow \boxed{\div 10} \longrightarrow \boxed{\div 5} \longrightarrow$?

$40 \longrightarrow \boxed{\div 5} \longrightarrow \boxed{\div 10} \longrightarrow$?

$600 \div 3{,}000 = \boxed{}$

$600 \longrightarrow \boxed{\div 3} \longrightarrow \boxed{\div 1{,}000} \longrightarrow$?

$600 \longrightarrow \boxed{\div 1{,}000} \longrightarrow \boxed{\div 3} \longrightarrow$?

b) Find the pairs of calculations that have the same answer.

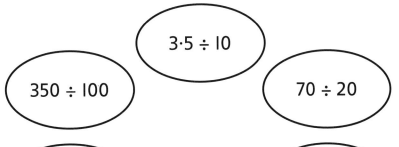

$3{\cdot}5 \div 10$

$350 \div 100$

$70 \div 20$

$7 \div 200$

$35 \div 1{,}000$

$70 \div 200$

To work out $70 \div 20$, I will divide by 2, then divide by 10.

15

Decimals as fractions

Discover

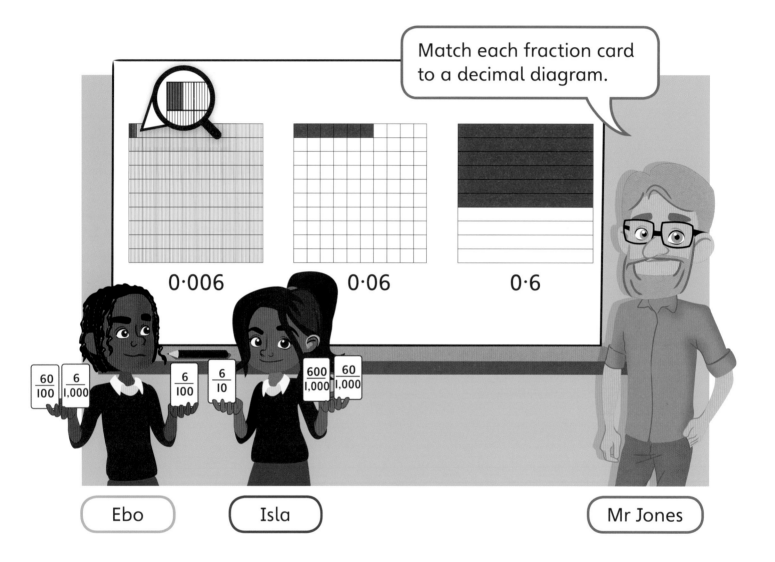

Match each fraction card to a decimal diagram.

0·006

0·06

0·6

$\frac{60}{100}$ $\frac{6}{1,000}$

$\frac{6}{100}$ $\frac{6}{10}$

$\frac{600}{1,000}$ $\frac{60}{1,000}$

Ebo

Isla

Mr Jones

1 **a)** Isla and Ebo are matching fraction cards to decimal diagrams. Which fractions are equivalent to each decimal?

b) Simplify the fractions on the cards, if possible.

Share

a) Use equivalent fractions to match the decimals.

O	•	Tth	Hth	Thth
0	•	0	0	6

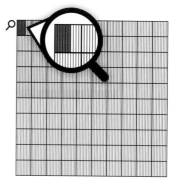

$\dfrac{6}{1,000}$

O	•	Tth	Hth	Thth
0	•	0	6	

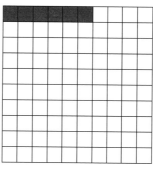

$\dfrac{6}{100} = \dfrac{60}{1,000}$

O	•	Tth	Hth	Thth
0	•	6		

$\dfrac{6}{10} = \dfrac{60}{100} = \dfrac{600}{1,000}$

$\dfrac{6}{1,000}$ is equivalent to 0·006.

$\dfrac{6}{100}$ and $\dfrac{60}{1,000}$ are equivalent to 0·06.

$\dfrac{6}{10}$, $\dfrac{60}{100}$ and $\dfrac{600}{1,000}$ are equivalent to 0·6.

b) $\dfrac{6}{1,000}$ can be simplified to $\dfrac{3}{500}$.

$\dfrac{6}{100}$ and $\dfrac{60}{1,000}$ can be simplified to $\dfrac{3}{50}$.

$\dfrac{6}{10}$, $\dfrac{60}{100}$ and $\dfrac{600}{1,000}$ can be simplified to $\dfrac{3}{5}$.

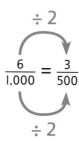

$\div 2$

$\dfrac{6}{1,000} = \dfrac{3}{500}$

$\div 2$

Think together

1 Write each decimal as a fraction. The first has been done for you.

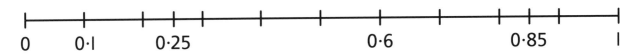

0   0·1   0·25   0·6   0·85   I

 $0·1 = \frac{1}{10}$ 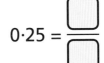 $0·25 = \frac{\square}{\square}$ 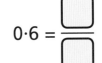 $0·6 = \frac{\square}{\square}$ 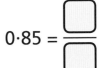 $0·85 = \frac{\square}{\square}$

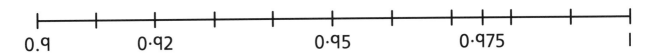

0.9   0·92   0·95   0·975   I

$0·92 = \frac{92}{100}$ $0·95 = \frac{\square}{\square}$ $0·975 = \frac{\square}{\square}$

2 Write each of these decimals as an improper fraction and as a mixed number.

2·3	
2	0·3

O	•	Tth	Hth
2	•	5	3

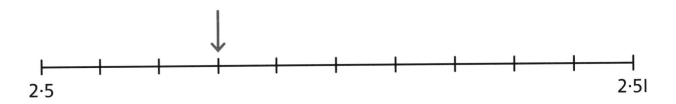

2·5   2·51

3 **a)** Complete the simplifications.

CHALLENGE

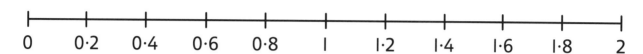

0   0·2   0·4   0·6   0·8   1   1·2   1·4   1·6   1·8   2

To simplify a fraction, you need to find a common factor of the numerator and the denominator.

$0.2 \longrightarrow \frac{2}{10} \longrightarrow \frac{1}{5}$

$1.2 \longrightarrow \frac{12}{10} \longrightarrow \boxed{}\frac{\boxed{}}{\boxed{}}$

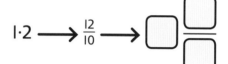

$0.4 \longrightarrow \frac{4}{10} \longrightarrow \frac{\boxed{}}{\boxed{}}$

$1.4 \longrightarrow \frac{\boxed{}}{\boxed{}} \longrightarrow \boxed{}\frac{\boxed{}}{\boxed{}}$

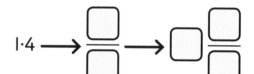

$0.6 \longrightarrow \frac{\boxed{}}{\boxed{}} \longrightarrow \frac{\boxed{}}{\boxed{}}$

$1.6 \longrightarrow \frac{\boxed{}}{\boxed{}} \longrightarrow \boxed{}\frac{\boxed{}}{\boxed{}}$

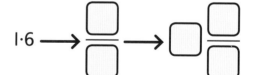

$0.8 \longrightarrow \frac{\boxed{}}{\boxed{}} \longrightarrow \frac{\boxed{}}{\boxed{}}$

$1.8 \longrightarrow \frac{\boxed{}}{\boxed{}} \longrightarrow \boxed{}\frac{\boxed{}}{\boxed{}}$

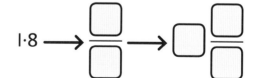
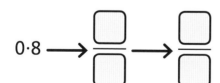
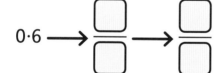
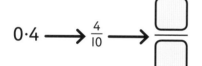

b) Convert these decimals into fractions and simplify them as far as you can.

| 0·25 | 0·125 | 0·875 | 0·35 | 0·95 |

19

Fractions as decimals ❶

Discover

❶ a) Sofia pours $\frac{1}{10}$ of a litre of liquid from the flask into beaker A.

She then pours $\frac{3}{4}$ of a litre of the liquid into beaker B.

If Sofia reads the scale of each beaker, what measurements will she record?

b) How much liquid is left in the flask?

Share

a) The scale on beaker A is in decimals.
Work out what $\frac{1}{10}$ of a litre is in decimals.

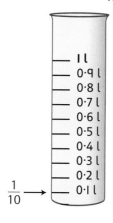

I used a place value grid to check my answer.

O	•	Tth	Hth
0	•	1	

$\frac{1}{10}$ is equivalent to 0·1.

Sofia will record 0·1 l for beaker A.

For beaker B, work out what $\frac{3}{4}$ of a litre is in decimals.

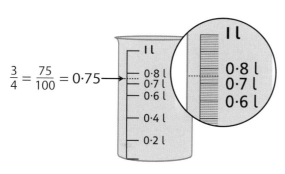

$\frac{3}{4} = \frac{75}{100} = 0.75$

I know that $\frac{1}{4}$ is 0·25, so $\frac{3}{4}$ is 0·75.

Sofia will record 0·75 l for beaker B.

b)

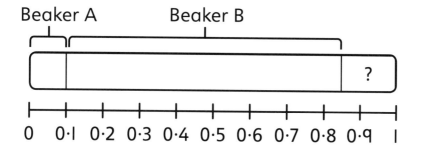

0·1 + 0·75 = 0·85

1 − 0·85 = 0·15

There is 0·15 l of liquid left in the flask.

Think together

1 What will $\frac{355}{1,000}$ kg look like on the display of this balance?

$\frac{355}{1,000}$ kg

?·??? kg

$\frac{355}{1,000}$		
☐	☐	☐
10	100	1,000

The display will show ☐·☐☐☐ kg.

2 Convert these fractions into decimals.

$\frac{6}{20}$ $\frac{7}{20}$ $\frac{16}{20}$ $\frac{17}{20}$

1									
$\frac{1}{10}$	$\frac{1}{10}$	$\frac{1}{10}$	$\frac{1}{10}$	$\frac{1}{10}$	$\frac{1}{10}$	$\frac{1}{10}$	$\frac{1}{10}$	$\frac{1}{10}$	$\frac{1}{10}$
$\frac{1}{20}$ $\frac{1}{20}$	$\frac{1}{20}$ $\frac{1}{20}$	$\frac{1}{20}$ $\frac{1}{20}$	$\frac{1}{20}$ $\frac{1}{20}$	$\frac{1}{20}$ $\frac{1}{20}$	$\frac{1}{20}$ $\frac{1}{20}$	$\frac{1}{20}$ $\frac{1}{20}$	$\frac{1}{20}$ $\frac{1}{20}$	$\frac{1}{20}$ $\frac{1}{20}$	$\frac{1}{20}$ $\frac{1}{20}$

0   0·1   0·2   0·3   0·4   0·5   0·6   0·7   0·8   0·9   1

× 5

$$\frac{7}{20} = \frac{\boxed{}}{100}$$

× 5

3 Convert these fractions to decimals and arrange them from smallest to largest.

$\frac{9}{10}$ $\frac{9}{20}$ $\frac{19}{10}$ $\frac{109}{100}$

$\frac{9}{50}$ $\frac{9}{25}$ $\frac{99}{1,000}$ $\frac{909}{10}$ $\frac{90}{250}$

I will use a part-whole model.

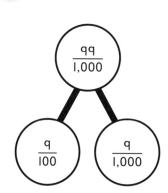

I will convert some of the fractions into mixed numbers.

For example, $\frac{109}{100} = 1\frac{9}{100}$.

_____ , _____ , _____ , _____ , _____ , _____ , _____ , _____ , _____

Smallest   Largest

23

Fractions as decimals ❷

1　**a)** What decimal number is the arrow pointing to?

　　b) Label all the decimals on the number line.

Share

a) The arrow points to $\frac{3}{8}$. Convert this fraction to a decimal.

Method 1

Find the fractions for quarters, then split each quarter into eighths.

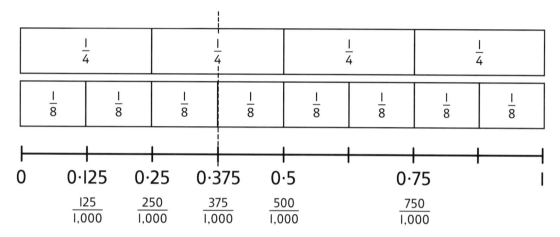

Then convert the fraction to a decimal.

$\frac{3}{8} = \frac{375}{1,000} = 0.375$

Method 2

$\frac{3}{8}$ can also be thought of as $3 \div 8$, so use division.

$$
\begin{array}{r}
0\ \cdot \\
\hline
8\,\big|\,3\ \cdot\ {}^{3}0
\end{array}
\qquad
\begin{array}{r}
0\ \cdot\ 3 \\
\hline
8\,\big|\,3\ \cdot\ {}^{3}0\ {}^{6}0
\end{array}
\qquad
\begin{array}{r}
0\ \cdot\ 3\ 7 \\
\hline
8\,\big|\,3\ \cdot\ {}^{3}0\ {}^{6}0\ {}^{4}0
\end{array}
\qquad
\begin{array}{r}
0\ \cdot\ 3\ 7\ 5 \\
\hline
8\,\big|\,3\ \cdot\ {}^{3}0\ {}^{6}0\ {}^{4}0
\end{array}
$$

$\frac{3}{8} = 3 \div 8 = 0.375$

The arrow is pointing to 0.375.

b) Each interval represents $\frac{1}{8}$. We can use equivalence with quarters to help us.

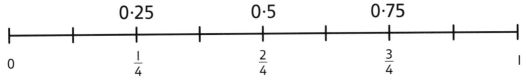

$\frac{2}{8} = \frac{1}{4} = 0 \cdot 25$ $0 \cdot 25 = \frac{25}{100}$ or $\frac{250}{1,000}$ So, $\frac{1}{8} = \frac{125}{1,000} = 0 \cdot 125$

Each interval increases by $0 \cdot 125$.

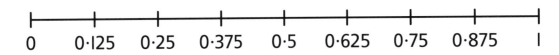

Think together

1 **a)** Danny is trying to convert $\frac{1}{3}$ to a decimal. Which of these is the correct way to start the division? Complete the correct division.

$$1\overline{)3 \cdot 0}^{\,3\,\cdot} \qquad 3\overline{)1 \cdot {}^{1}0}^{\,0\,\cdot}$$

Numbers that have a repeating decimal are called **recurring decimals**.

b) What do you notice?

2 Write each of these calculations as a fraction and a decimal.

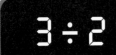 $3 \div 2$ 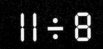 $11 \div 8$ 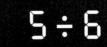 $5 \div 6$ $2 \div 3$

3 Discuss different methods for converting these fractions into decimals.

$$\frac{6}{9} \qquad \frac{3}{25} \qquad \frac{3}{50}$$

$$\frac{5}{12} \qquad \frac{300}{450} \qquad \frac{6}{90}$$

Convert each fraction into a decimal using an efficient method.

Round any recurring decimals to three decimal places.

The phrase 'three decimal places' can also be shortened to 'three dp', which is quicker to write and say.

To round to three dp, I will look at the fourth decimal place and decide whether to round up or down.

27

→ **Practice book 6B p18**

Multiplying decimals ❶

Discover

❶ **a)** What is the total volume of the 3 drinks cans?

b) What is the total volume of 30 cans?

Share

a) Work out 3 × 0·3.

Method 1: Use known facts.

3 × 3 = 9

3 × 0·3 = 0·9

T	O	•	Tth
		•	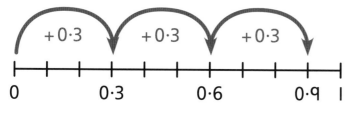

Method 2: Use fractions.

$0·3 = \frac{3}{10}$

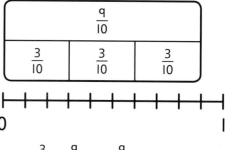

$3 × \frac{3}{10} = \frac{9}{10}$ $\frac{9}{10} = 0·9$

Method 3: Count in decimal steps.

0·3 + 0·3 + 0·3 = 0·9

Method 4: Convert the measuring units.

0·3 l = 300 ml

3 × 300 ml = 900 ml

900 ml = 0·9 l

The total volume of the 3 drinks cans is 0·9 litres.

b) 30 × 0·3 = ?

10 × 0·3 = 3, so the volume of 10 cans is 3 litres.

There are 3 groups of 10 cans.

3 groups of 3 litres is 9 litres.

The total volume of 30 cans is 9 litres.

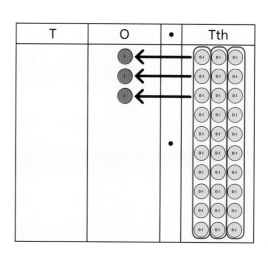

Think together

1 Write and complete a multiplication for each model.

a)

T	O	•	Tth
		•	(0·1) (0·1) (0·1) (0·1) (0·1) (0·1)

c)

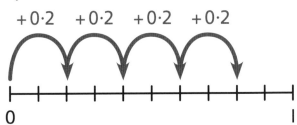

b)

T	O	•	Tth	Hth
		•		(0·01) (0·01) (0·01) (0·01) (0·01) (0·01) (0·01) (0·01) (0·01)

d)

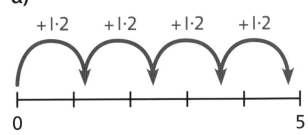

2 Predict which multiplication will give the greatest and the smallest **products**.

Complete each calculation to check.

The answer to a multiplication is called the product.

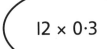

12 × 0·3

0·02 × 32

21 × 0·04

3 **a)** Complete these sets of multiplications.

$2 \times 3 = 6$ $20 \times 3 = \boxed{}$

$0{\cdot}2 \times 3 = 0{\cdot}6$ $20 \times 0{\cdot}3 = \boxed{}$

$0{\cdot}02 \times 3 = \boxed{}$ $20 \times 0{\cdot}03 = \boxed{}$

$2 \times 0{\cdot}3 = \boxed{}$ $200 \times 0{\cdot}3 = \boxed{}$

$2 \times 0{\cdot}03 = \boxed{}$ $2{,}000 \times 0{\cdot}03 = \boxed{}$

b) Explain how the answers are related.

c) Explain why some of the calculations have the same answer.

> I will use a place value grid to explore how the answers are related.

	H	T	O	•	Tth	Hth
2×3			6	•		
$0{\cdot}2 \times 3$			0	•	6	
$0{\cdot}02 \times 3$				•		

→ **Practice book 6B p21**

Multiplying decimals ❷

Discover

1 **a)** How tall is the fence panel?

b) The panel is 1·2 m wide. The whole fence will be made of 6 panels. How long will the whole fence be?

Share

a) There are 6 planks that are each 0·3 m high.

Work out 6 × 0·3.

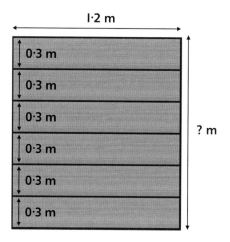

Method 1

Use known facts and exchange.

I know 6 × 3 is 18. So 6 groups of 3 tenths is 18 tenths. Now I will exchange.

T	O	•	Tth

6 × 0·3 = 1·8

Method 2

Use fractions and convert improper fractions.

$0·3 = \frac{3}{10}$

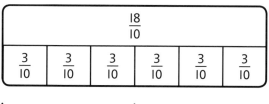

$6 \times \frac{3}{10} = \frac{18}{10} = 1\frac{8}{10} = 1·8$

The fence panel is 1·8 m tall.

b) There are 6 panels that are each 1·2 m long, so work out 6 × 1·2.

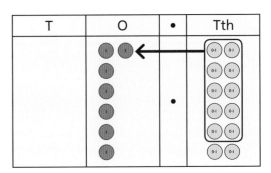

T	O	•	Tth

$6 \times 1 = 6$ $6 \times 0·2 = 1·2$ $6 + 1·2 = 7·2$

The whole fence will be 7·2 m long.

Think together

1 Write and complete a multiplication for each of these representations.

a)

(0·1) (0·1) (0·1) (0·1)
(0·1) (0·1) (0·1) (0·1)
(0·1) (0·1) (0·1) (0·1)

☐ × ☐ = ☐

b)

⬤ ⬤ ⬤ ⬤

←→←→←→←→
1·3 cm 1·3 cm 1·3 cm 1·3 cm

☐ × ☐ = ☐

2 Use the known fact to calculate the other multiplications.

a) 1·8 × 4 = ☐

c) 180 × 0·4 = ☐

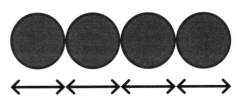

18 × 4 = 72

b) 18 × 0·4 = ☐

d) 18 × 0·04 = ☐

3 **a)** Calculate the area of each rectangle.

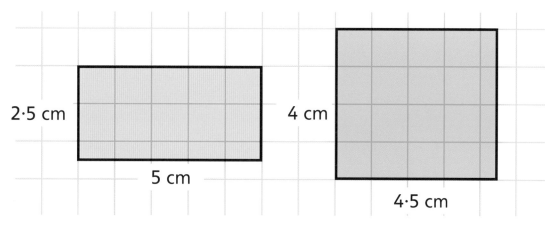

2·5 cm

5 cm

4 cm

4·5 cm

◻ × ◻ = ◻   ◻ × ◻ = ◻

b) Calculate the areas of these rectangles.

15 m

0·03 m

1·2 m

5 m

0·09 m

12 m

◻ × ◻ = ◻   ◻ × ◻ = ◻   ◻ × ◻ = ◻

I will use an area model and partition to simplify each calculation.

5 m

1 m

0·2 m

→ Practice book 6B p24

Dividing decimals ①

Discover

① **a)** Four small blocks balance a 0·8 kg box. What is the mass of each block?

b) How many blocks will balance an 8 kg crate?

Share

a)

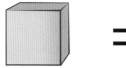 =

The total mass of the 4 blocks is 0·8 kg, so find 0·8 ÷ 4.

I think I can use multiplication facts to help me.

0·8 ÷ 4 = ?

4 × ? = 0·8

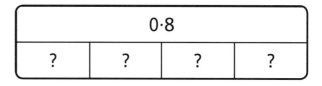

0·8			
?	?	?	?

4 × 2 = 8

So, 4 × 0·2 = 0·8

8 ÷ 4 = 2

0·8 ÷ 4 = 0·2

I think I can solve this by using sharing to find out what 0·8 is when it is shared into four parts.

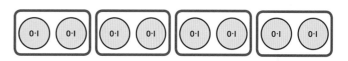

The mass of each block is 0·2 kg.

I will check my answer with multiplication.

40 × 0·2 = 4 × 2

40 × 0·2 = 8

b) 8 kg is ten times as heavy as 0·8 kg.

So, 10 times as many blocks will balance the scale.

4 × 10 = 40

40 blocks will balance an 8 kg crate.

Think together

1 A bottle contains 1·2 litres of juice. It is shared equally between 3 glasses. How many litres of juice are in each glass?

T	O	•	Tth	Hth
	1	•	0·1 0·1	

I will exchange a 1 for 10 tenths.

There are ☐ litres of juice in each glass.

2 Explain how each of these models can be used to solve 4·8 ÷ 6 = ☐ .

T	O	•	Tth	Hth
	1 →		0·1 0·1	
	1 →		0·1 0·1	
	1 →	•	0·1 0·1	
	1 →		0·1 0·1	

I know 48 ÷ 6 = 8, so …

Danny

4·8

Hmm,
6 × 0·5 = 3 …
6 × 0·6 = 3·6 …

Bella

CHALLENGE

3 All three of these towers are 14·4 cm tall.

What is the height and width of one red block?

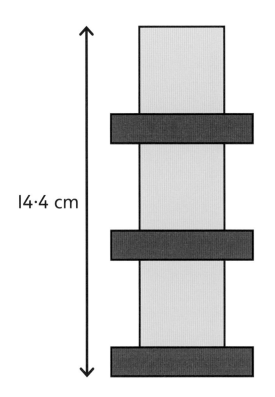

14·4 cm

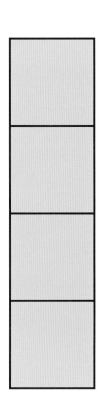

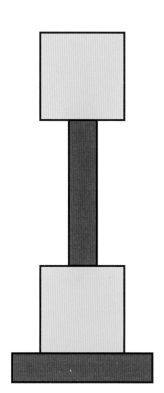

I can see an easy way to find the height of one yellow block.

A red block is ☐ wide and ☐ long.

39

Dividing decimals ②

Discover

4·24 m

? m

Toshi

1 **a)** Calculate the width of one paving slab.

b) How many slabs would Toshi need to make a path that is at least 10 m long?

Share

a) There are 8 slabs, so calculate 4·24 ÷ 8.

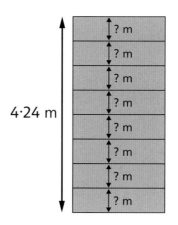

4·24 m

I cannot recognise any multiplication facts to help me solve this. I will try exchanging for hundredths.

4·24 is 424 hundredths.

$$8 \overline{\smash{\big)}\, 4\ ^42\ ^24} \quad \begin{array}{c} 5\ \ 3 \end{array}$$

424 cm ÷ 8 = 53 cm

So, 4·24 ÷ 8 is 53 hundredths, or 0·53 m.

0·53 m = 53 cm

Short division can be used with decimals. Remember the decimal points must be aligned.

$$8 \overline{\smash{\big)}\, 4\ \cdot\ 2\ \ 4} \quad \begin{array}{c} \cdot \end{array}$$

$$8 \overline{\smash{\big)}\, 4\ \cdot\ ^42\ \ 4} \quad \begin{array}{c} 0\ \cdot \end{array}$$

$$8 \overline{\smash{\big)}\, 4\ \cdot\ ^42\ ^24} \quad \begin{array}{c} 0\ \cdot\ 5 \end{array}$$

$$8 \overline{\smash{\big)}\, 4\ \cdot\ ^42\ ^24} \quad \begin{array}{c} 0\ \cdot\ 5\ \ 3 \end{array}$$

4·24 ÷ 8 = 0·53

One paving slab is 0·53 m or 53 cm wide.

b) Find out how many 0·53 m slabs will be needed to make a path that is at least 10 m long.

10 m = 1,000 cm

10 × 53 = 530

20 × 53 = 1,060

19 × 53 = 1,007

18 × 53 = 954

1,060		
1,007		53
954	53	53

Toshi would need 19 slabs to make a path that is at least 10 m long.

Think together

1 A gardener cuts a 3·5 m length of wood into 4 equal planks.
Complete the division to calculate the length of each plank.

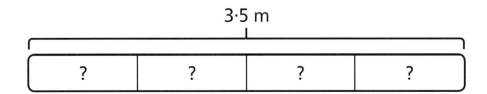

3·5 m

| ? | ? | ? | ? |

3 · 5

2 Predict whether the answers to these divisions will have one or two decimal places. Then complete the divisions to check.

a) 0·8 ÷ 5 = ☐

b) 85·8 ÷ 5 = ☐

c) 8·05 ÷ 5 = ☐

d) 508·5 ÷ 5 = ☐

3 **a)** Estimate the answer to each of the divisions by choosing one of these options.

| Less than 0·1 | Between 0·1 and 1 | Between 1 and 10 | More than 10 |

3·2 ÷ 4

9·9 ÷ 11

32 ÷ 5

I will approximate by rounding.

100·4 ÷ 8

1,234 ÷ 4

0·8 ÷ 80

Choose an appropriate method to complete each division and check your estimates.

b) Which numbers complete each calculation?

$0·16 ÷ \boxed{} = 0·04$

$\boxed{} ÷ 50 = 0·24$

$1·2 = 24 ÷ \boxed{}$

$0·25 = \boxed{} ÷ 4$

43

→ Practice book 6B p30

End of unit check

1 Which calculation is equivalent to 32 ÷ 1,000?

A 0·32 ÷ 10   **B** 0·032 × 10   **C** 320 ÷ 100   **D** 3,200 ÷ 1,000

2 What is this fraction as a decimal?

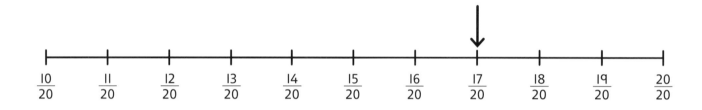

A 0·17   **B** 0·34   **C** 0·85   **D** 17·20

3 Which calculation does not accurately match the diagram?

0·01	0·01	0·01	0·01	0·01
0·01	0·01	0·01	0·01	0·01
0·01	0·01	0·01	0·01	0·01

A 15 × 0·01 = 0·15   **C** 5 × 0·03 = 0·15

B 3 × 0·5 = 0·15   **D** 0·05 × 3 = 0·15

4 Which division contains an error?

A
$$5 \overline{)1 \cdot {}^12 \; {}^25}$$
quotient: $0 \cdot 2 \; 5$

C
$$6 \overline{)1 \cdot {}^15 \; 3}$$
quotient: $0 \cdot 2 \; 3$

B
$$8 \overline{)3 \; 3 \cdot {}^12 \; {}^48}$$
quotient: $4 \cdot 1 \; 6$

D
$$8 \overline{)1 \cdot {}^12 \; {}^40}$$
quotient: $0 \cdot 1 \; 5$

5 What completes $\boxed{} \div 12 = 0 \cdot 8$?

A $0 \cdot 96$ **B** 96 **C** $9 \cdot 6$ **D** $9 \cdot 06$

6 Fill in the missing number.

$3 \cdot 5 \times 7 = 28 \times \boxed{}$

45

→ Practice book 6B p33

Unit 8
Percentages

In this unit we will …

⚡ Develop a deeper understanding of percentages as parts of 100

⚡ Understand a range of methods to work out percentages

⚡ Find 1% and multiples of 1%

⚡ Work out missing values, such as 30% of ? = 60

⚡ Convert, order and solve problems involving fractions, percentages and decimals

Do you remember what this model is called?

It can be used to represent percentages of amounts and to solve problems.

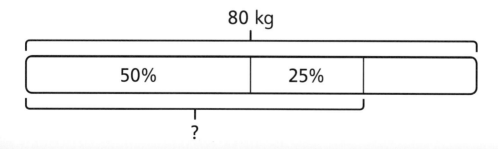

80 kg

| 50% | 25% | |

?

We will need some maths words.
Do you know what they all mean?

per cent (%)   percentage

parts   whole   decimal   fraction

divide   share   multiply   convert

compare   order   equivalent fraction

simplify   less than (<)   greater than (>)

We will need to use a number line too.

You can use this to help you to order decimals, fractions and percentages.

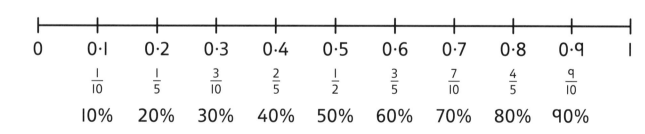

0	0·1	0·2	0·3	0·4	0·5	0·6	0·7	0·8	0·9	1
	$\frac{1}{10}$	$\frac{1}{5}$	$\frac{3}{10}$	$\frac{2}{5}$	$\frac{1}{2}$	$\frac{3}{5}$	$\frac{7}{10}$	$\frac{4}{5}$	$\frac{9}{10}$	
	10%	20%	30%	40%	50%	60%	70%	80%	90%	

Percentage of

Discover

Paint 50% of the circles yellow and paint 25% blue.

Mr Jones

Sofia

1 a) How many circles should Sofia paint yellow?

   b) Show two different ways she could paint the blue circles.

Share

a) Each painting has 100 circles. There are 4 paintings.

I know that 50% means 50 out of 100, or $\frac{50}{100}$. I will paint 50 out of every 100 circles.

I know that $\frac{50}{100}$ is equivalent to $\frac{1}{2}$. So I will find half of the total.

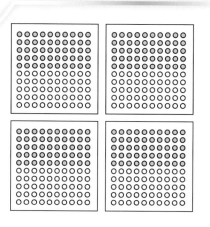
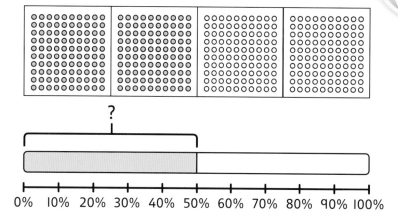

50 + 50 + 50 + 50 = 200

50% of 400 is 200.

200 circles are painted yellow.

There are 400 circles in total.

$\frac{1}{2}$ of 400 is 200.

200 circles are painted yellow.

Sofia should paint 200 circles yellow.

b) 25% is 25 out of every 100.

$$\frac{25}{100} = \frac{1}{4}$$

Sofia could paint the circles blue in these two ways.

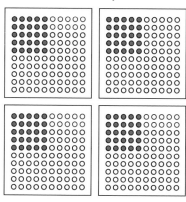
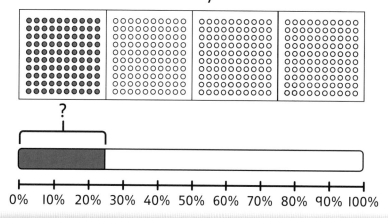

Think together

1 300 people visit the fair.

50% win a toffee apple. 25% win a balloon. 10% win a teddy.

How many people win each prize?

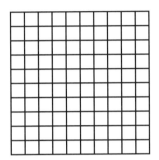 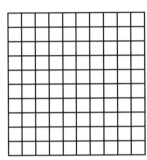 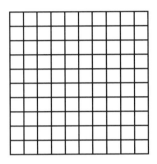

I know that 25% means 25 out of every 100 and that 10% means 10 out of every 100.

50% of 300 is ☐ . So, ☐ people win a toffee apple.

25% of 300 is ☐ . So, ☐ people win a balloon.

10% of 300 is ☐ . So, ☐ people win a teddy.

2 Copy the grid. Shade 50% yellow, 25% red and 10% blue.

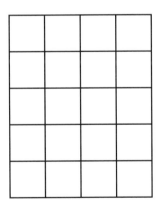

20 squares

50%	50%

25%	25%	25%	25%

10%	10%	10%	10%	10%	10%	10%	10%	10%	10%

3 Bella wants to buy a pair of trainers and a skateboard in the sale.

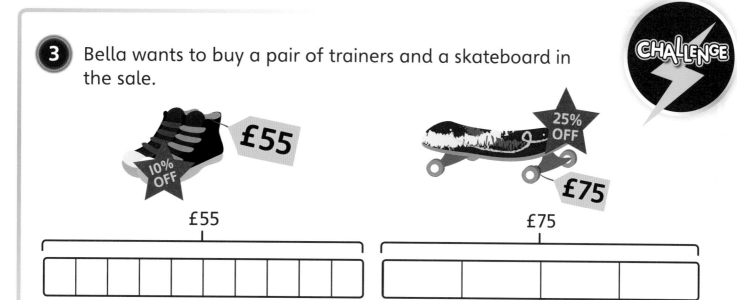

£55

£75

CHALLENGE

First, I can work out 50% of 75.
Then I can use that to find 25%.

a) How much are the trainers reduced by? Once reduced, how much do they cost?

b) How much is the skateboard reduced by? Once reduced, how much does it cost?

c) Explain Bella's mistake using equipment, diagrams and words.

To find 10% I can divide by 10.
So, to find 25% I can divide by 25.

Bella

51

→ Practice book 6B p35

Percentage of ❷

Discover

① **a)** How many motorbikes are on the ferry?

b) 25% of the vehicles on the ferry are vans.

How many more vans than motorbikes are there?

Share

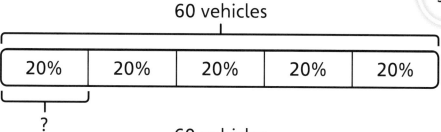

There are 5 equal parts of 20% in 100%, so I divided by 5.

a)

60 vehicles

| 20% | 20% | 20% | 20% | 20% |

?

60 vehicles

| 10% | 10% | 10% | 10% | 10% | 10% | 10% | 10% | 10% | 10% |

?

I found 10% of the total and then doubled it to find 20%.

Method I

$\frac{20}{100} = \frac{1}{5}$

$60 \div 5 = 12$

Method 2

20% is 2 × 10%.

10% of 60 is 6.

So, 20% of 60 is 12.

There are 12 motorbikes on the ferry.

b) 25% of 60 is 15. There are 15 vans on the ferry.

There are 12 motorbikes on the ferry.

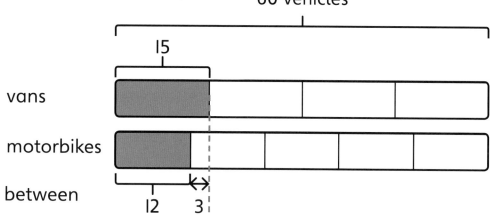

60 vehicles

15

vans

motorbikes

12   3

The difference between 12 and 15 is 3.

There are 3 more vans than motorbikes on the ferry.

Think together

1 **a)** Cover 10% of the grid with cubes or counters.

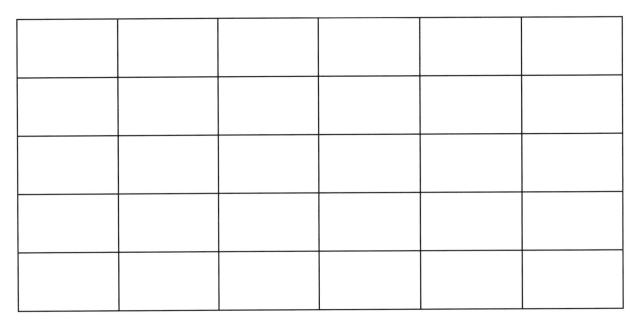

b) Now cover 20% of the grid with cubes or counters.

2 Work out these calculations. Discuss how each pair of calculations is related.

a) 10% of 120 = ☐

20% of 120 = ☐

b) 20% of 150 = ☐

20% of 75 = ☐

c) 20% of 80 = ☐

10% of ☐ = 16

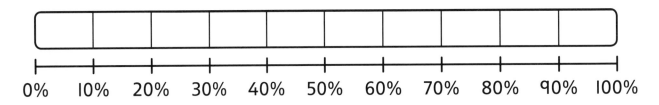

0%　10%　20%　30%　40%　50%　60%　70%　80%　90%　100%

I can use the bar model to help me to explain.

3 Ebo is solving $240 \div 5 = \boxed{}$.

I can use percentages to solve this mentally.

10% of 240 is 24.

20% of 240 is 48.

So, $240 \div 5 = 48$.

Ebo

a) Do you agree with Ebo's answer?

How does his method work?

b) Does Ebo's method work when dividing any number by 5?

Try these:

$320 \div 5 = \boxed{}$

$70 \div 5 = \boxed{}$

$55 \div 5 = \boxed{}$

$12 \div 5 = \boxed{}$

I wonder if Ebo's method is always the most efficient.

55

Percentage of ③

Discover

1% of the chocolate bars in each box have a winning rainbow ticket!

2,500 bars

500 bars

200 bars

1,000 bars

WINNER!

① **a)** In a box of 500 chocolate bars, how many bars have a winning rainbow ticket?

b) How many rainbow tickets are in each of the other boxes?

Share

a)

1% is $\frac{1}{100}$.

I found 1 out of every 100, but I knew I could also divide the whole by 100.

1% of 500 = 1 + 1 + 1 + 1 + 1 = 5

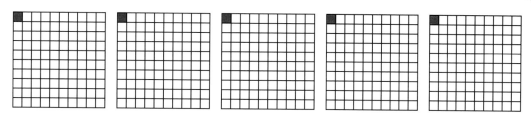

	Th	H	T	O
Whole amount		5	0	0
$\frac{1}{100}$ of the amount				5

I used a place value grid to divide the whole by 100.

In a box of 500 chocolate bars, 5 bars have rainbow tickets.

b) 1% is always $\frac{1}{100}$ of the whole amount.

Whole number	$\frac{1}{100}$ of the number	1% of the number
200	200 ÷ 100 = 2	1% of 200 is 2
1,000	1,000 ÷ 100 = 10	1% of 1,000 is 10
2,500	2,500 ÷ 100 = 25	1% of 2,500 is 25

In a box of 200 bars, there are 2 rainbow tickets.

In a box of 1,000 bars, there are 10 rainbow tickets.

In a box of 2,500 bars, there are 25 rainbow tickets.

Think together

1

300 bars

500 bars

1,500 bars

4% of the chocolate bars in each box have a silver ticket.

Use these examples to help you to work out how many silver tickets there are in each box.

	Th	H	T	O
Whole amount		3	0	0
1% of the amount				3

Whole amount	1% of the whole amount	4% of the whole amount
300		
500		
1,500		

2 Complete the percentages.

a) 100% is 150 cm

10% is ☐ cm

1% is ☐ cm

2% is ☐ × 2 = ☐ cm

b) 100% is 24 m

10% is ☐ m

1% is ☐ m

3% is ☐ × ☐ = ☐ m

3 Emma and Isla are discussing their knowledge of percentages.

> Now I can use division to find:
> 50%
> 25%
> 20%
> 10% and
> 1% of any number.

Emma

> I think I can work out any percentage. First, I find 1% then I multiply. For example, to find 20%, I find 1% then multiply by 20.

Isla

> I think that sometimes Emma's method is more efficient and sometimes Isla's is.

Complete the table.

	Emma's method	Isla's method	Percentage of £35
50%	Divide the whole by 2.	Find 1% and then multiply by 50.	£17·50
25%	Divide …	Find …	
20%			
10%			

59

Percentage of ❹

Discover

I **a)** What was the weight of paper that Class I recycled?

b) In weight, how much more paper than plastic did Class I recycle?

Share

a) Class I recycled a total weight of 80 kg. The bar chart shows that 75% of the weight was paper.

I will work out 1% of 80 and then multiply it by 75.

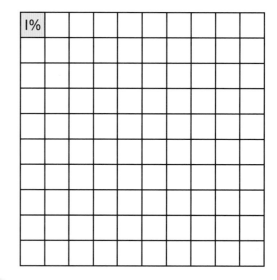

$80 \div 100 = 0.8$

$$
\begin{array}{r}
7\ 5 \\
\times \quad 0\cdot 8 \\
\hline
6\ \ 0\cdot 0 \\
\ {}_{4}
\end{array}
$$

I know that 75% is equal to $\frac{3}{4}$.

$80 \div 4 = 20$

$20 \times 3 = 60$

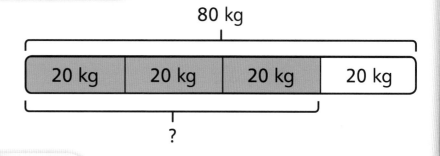

I will add 50% and 25%.

$\frac{1}{2}$ of $80 = 40$

$\frac{1}{4}$ of $80 = 20$

$40 + 20 = 60$

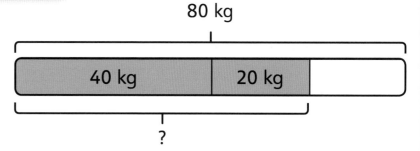

75% of 80 kg is 60 kg. Class I recycled 60 kg of paper.

b) 100% – 75% = 25%

The percentage of plastic recycled was 25%.

25% of 80 kg is 20 kg.

Class I recycled 20 kg of plastic.

60 kg – 20 kg = 40 kg

Class I recycled 40 kg more paper than plastic.

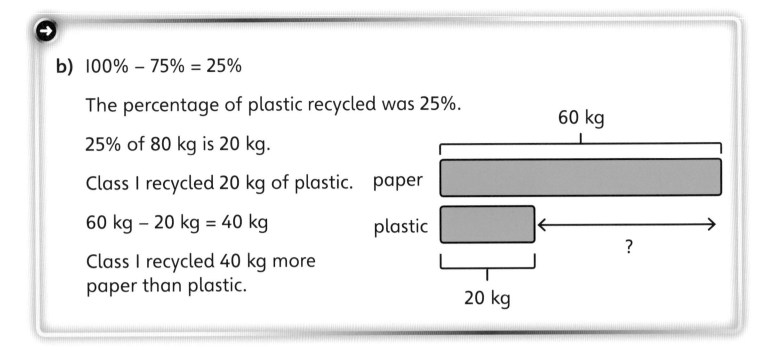

Think together

① Class 2 collected 120 kg of waste to recycle. 60% was paper and 40% was plastic.

The bar models show two different strategies. Use them to calculate the weight of paper and plastic that Class 2 recycled.

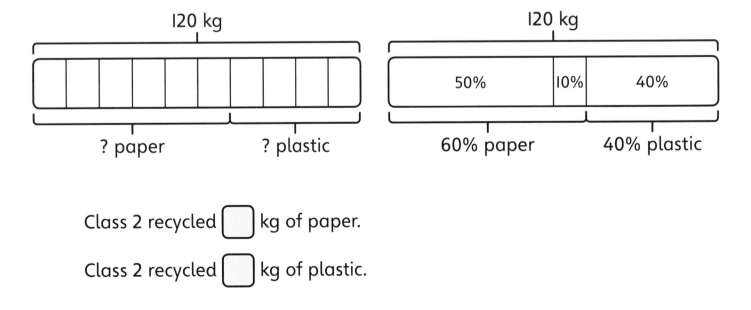

Class 2 recycled ☐ kg of paper.

Class 2 recycled ☐ kg of plastic.

2 Calculate these values.

5% of £300 = £ ⬚

15% of 300 cm = ⬚ cm

55% of 300 kg = ⬚ kg

95% of 30 km = ⬚ km

300

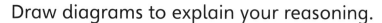

3 **a)** Discuss different strategies for finding these percentages of 320.

Draw diagrams to explain your reasoning.

CHALLENGE

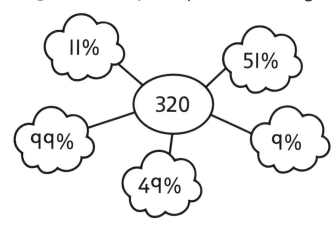

11%

51%

320

99%

9%

49%

b) A lettuce weighs 148 g. 95% of the weight is water.

What weight is not water?

I will find 95% of 148 and then subtract it from the whole.

I wonder if there is a more efficient method.

63

Finding missing values

Discover

1 a) What is the world record for women's long jump?

b) Lee jumps 49 cm in the high jump. This is 20% of the world record for men's high jump. What is the world record?

Share

a)

I think I have to work out 50% of 3·75 m.

I think you need to do it differently. In this question, I know a part and I have to work out the whole.

50% of ⬚ = 3·75

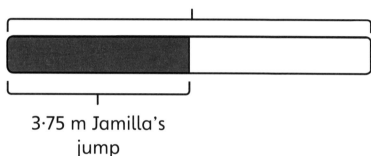

world record ?

3·75 m Jamilla's jump

3·75 × 2 = 7·5

The world record for women's long jump is 7·5 m.

b) 20% of ⬚ = 49 cm

20% is equivalent to $\frac{1}{5}$.

49 × 5 = 245

world record ?

49 cm Lee's jump

The world record for men's high jump is 245 cm.

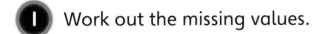

Think together

1 Work out the missing values.

a) 10% of ⬜ = 3   10% is equivalent to $\frac{1}{10}$ 3 × ⬜ = ⬜

?

| 3 | | | | | | | | | |

b) 20% of ⬜ = 5   5 × ⬜ = ⬜

?

| | | | | |

c) 25% of ⬜ = 30

| |

2 120 people arrive by bus to watch sports day. This is 40% of the spectators.

How many spectators are there in total?

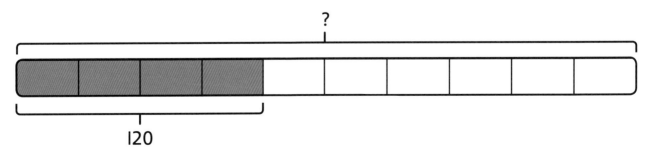

120

3 Find the solutions.

What is the same and what is different about how you found each solution?

a)

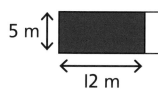

Guess my number.

Is it 30?

No, but that is exactly 15% of my number.

b) Jen has completed 30% of her 120 mile journey. How far has she travelled?

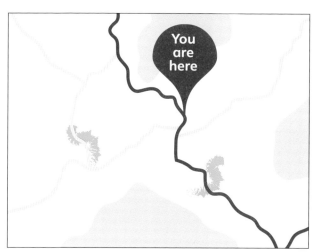

You are here

c) 15% of the rectangle is shaded.

5 m

12 m

What is the area of the whole rectangle?

5 m + 12 m = 17 m, so 15% of the rectangle must be 17 m. Now I just need to work out the whole.

I think I need to multiply to find the area. So, to find 15% of the rectangle, first I need to work out 5 m × 12 m.

→ Practice book 6B p47

Converting fractions to percentages

Discover

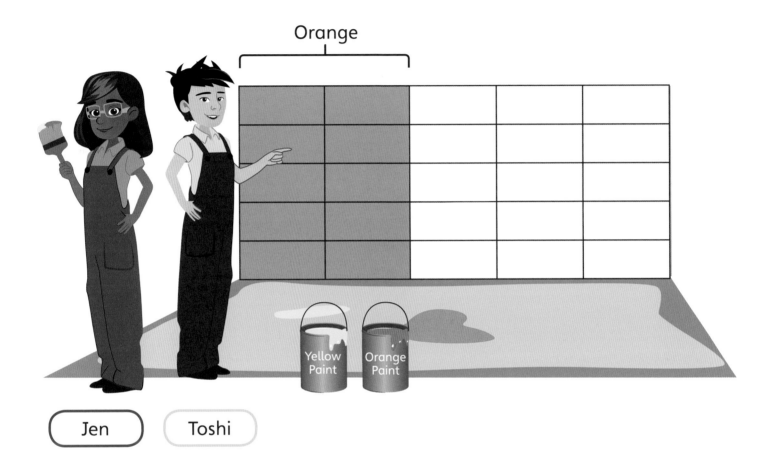

1 **a)** What percentage of the wall is painted orange?

b) Jen and Toshi want to paint 30% of the wall yellow.

How many rectangles do they need to paint yellow?

Share

a) The whole wall is 100%.

There are 25 rectangles on the wall.

$100 \div 25 = 4$

4%	4%	4%	4%	4%
4%	4%	4%	4%	4%
4%	4%	4%	4%	4%
4%	4%	4%	4%	4%
4%	4%	4%	4%	4%

Each rectangle represents 4% of the whole wall.

10 rectangles are painted orange.

$10 \times 4 = 40$

40% of the wall is painted orange.

> I found an equivalent fraction with 100 as the denominator.

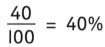

$$\frac{10}{25} = \frac{40}{100}$$

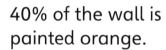

$$\frac{40}{100} = 40\%$$

40% of the wall is painted orange.

---

b) The whole wall has 25 rectangles. We need to find 30% of 25.

$25 \div 10 = 2 \cdot 5$

So, $10\% = 2 \cdot 5$

$3 \times 2 \cdot 5 = 7 \cdot 5$

O	O		Y	Y
O	O		Y	Y
O	O		Y	Y
O	O			Y
O	O			Y

Jen and Toshi need to paint 7·5 rectangles yellow.

69

Think together

1   What percentage of each grid is shaded?

a)

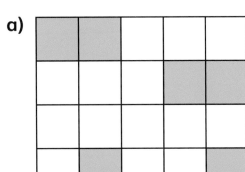

$$\frac{6}{20} = \frac{\boxed{}}{100}$$

$$\frac{\boxed{}}{100} = \boxed{} \%$$

$\boxed{}$ % of the grid is shaded.

b)

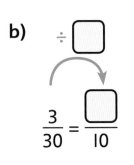

$$\div \boxed{}$$

$$\frac{3}{30} = \frac{\boxed{}}{10}$$

$$\times 10$$

$$\frac{\boxed{}}{10} = \frac{10}{\boxed{}}$$

$$\frac{\boxed{}}{100} = \boxed{} \%$$

$\boxed{}$ % of the grid is shaded.

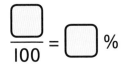

2   Here are the results of a survey.

Class 4

Yes	13
No	10
Don't know	2

Class 5

Yes	15
No	9
Don't know	6

Staff

Yes	2
No	17
Don't know	1

a) What percentage of each group voted 'Yes'?

b) What percentage of each group voted 'No'?

c) What percentage of each group voted 'Don't know'?

3 **a)** Amelia is comparing fractions. Help Amelia to convert the fractions into percentages.

Amelia

> It is more efficient if I change fractions into percentages to compare them.

 $\dfrac{8}{40}$

 $\dfrac{8}{400}$

 $\dfrac{45}{150}$

> Some of the denominators are greater than 100.
> I wonder what I can do.

> I will multiply and divide the fractions so that the denominator is 100.

b) Is it easier to compare percentages or fractions?

Explain your answer.

→ Practice book 6B p50

Equivalent fractions, decimals and percentages ❶

Discover

Where should we place each amount on the number line?

0   1

70%   0·2 $\frac{2}{5}$

Ambika

Zac

Miss Hall

❶ a) Help Ambika and Zac to place the amounts on the number line.

b) Where would $\frac{19}{20}$ go on the number line?

Share

Each fraction is in its simplest form.

a) Decimals, fractions and percentages can all represent equivalent numbers or amounts.

Ambika and Zac should place the numbers as shown below.

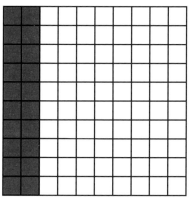
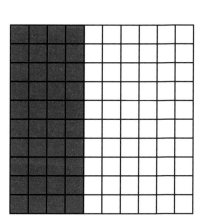
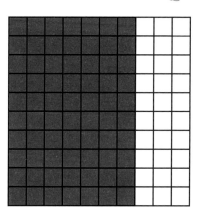

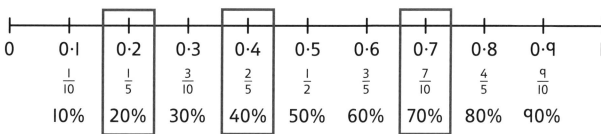

0	0·1	0·2	0·3	0·4	0·5	0·6	0·7	0·8	0·9	1
	$\frac{1}{10}$	$\frac{1}{5}$	$\frac{3}{10}$	$\frac{2}{5}$	$\frac{1}{2}$	$\frac{3}{5}$	$\frac{7}{10}$	$\frac{4}{5}$	$\frac{9}{10}$	
	10%	20%	30%	40%	50%	60%	70%	80%	90%	

b) $\frac{19}{20}$ is equivalent to $\frac{95}{100}$.

$\frac{19}{20} = 0.95 = 95\%$

$\frac{19}{20}$ should be placed as shown below.

$$\times 5$$
$$\frac{19}{20} = \frac{95}{100}$$
$$\times 5$$

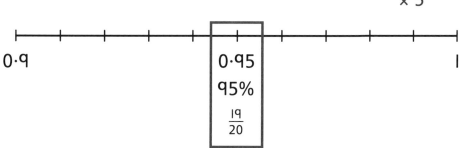

0·9					0·95					1
					95%					
					$\frac{19}{20}$					

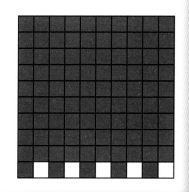

73

Think together

1 Which is the odd one out? Explain your answer.

$$0·1 \qquad \frac{2}{20} \qquad 10\% \qquad \frac{1}{10} \qquad 0·01 \qquad \frac{10}{100}$$

2 Copy and complete the table.

Decimal	Percentage	Fraction
0·5	50%	$\frac{1}{2}$
0·65		
	6%	
		$\frac{3}{10}$
0·07		

Remember to write each fraction in its simplest form.

3 Find the answers to each pair of calculations.

Discuss what you notice with a partner.

a) 50% of 3 = ☐

0·5 × 3 = ☐

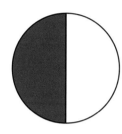 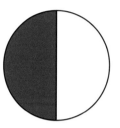 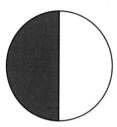

b) 10% of 3 = ☐

0·1 × 3 = ☐

c) 25% of 3 = ☐

3 × 0·25 = ☐

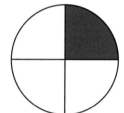 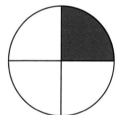 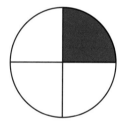

d) Try a few examples of your own, using different numbers.

e) Can you extend your findings further?

So, 20% is equivalent to 0·2.

I wonder how I could go about multiplying any number by 0·2.

I know how to find 5% of a number. I think I can use that to multiply by.

75

Equivalent fractions, decimals and percentages ❷

Discover

I scored 5 out of 20.

I got 6 out of 25.

Lee

Kate

1 **a)** Write Lee's score and Kate's score as fractions. Who was more accurate at the coconut shy?

b) Write a fraction that can be found between both scores.

Share

a) Lee's score is $\frac{5}{20}$. Kate's score is $\frac{6}{25}$.
Compare the two fractions $\frac{5}{20}$ and $\frac{6}{25}$.

> I know that $\frac{5}{20}$ is equivalent to $\frac{1}{4}$, but I cannot simplify $\frac{6}{25}$.

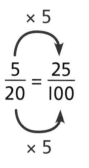
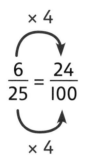

$$\frac{5}{20} \overset{\times 5}{\underset{\times 5}{=}} \frac{25}{100} \qquad \frac{6}{25} \overset{\times 4}{\underset{\times 4}{=}} \frac{24}{100}$$

> Converting fractions into percentages is a good way to compare them.

5 out of 20 is 25%. 6 out of 25 is 24%.

Lee was more accurate at the coconut shy.

b) To find a fraction between 24% and 25%, use equivalent fractions.

24% $\qquad$ 25%
$\frac{24}{100}$ $\qquad$ $\frac{25}{100}$
$\frac{240}{1,000}$ $\qquad$ $\frac{247}{1,000}$ $\qquad$ $\frac{250}{1,000}$

> I chose $\frac{247}{1,000}$. I can also write this as a decimal: 0·247.

Any fraction between $\frac{240}{1,000}$ and $\frac{250}{1,000}$ can be found between Lee's and Kate's scores, for example $\frac{247}{1,000}$.

> I wonder if I can find equivalent fractions for 24% and 25% other than thousandths.

77

Think together

1 Place these amounts in order.

| 55% | $\frac{5}{5}$ | 0·5 | 5% | $\frac{5}{25}$ | 0·505 |

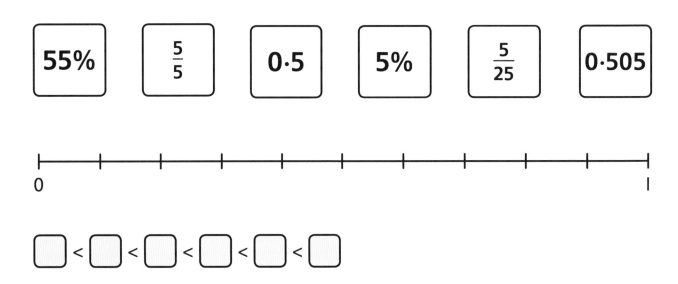

0   1

☐ < ☐ < ☐ < ☐ < ☐ < ☐

2 Once reduced, what is the difference in price between the most expensive and least expensive pair of trainers?

£47 $\frac{1}{4}$ OFF

£75·50   REDUCED BY 40%

£69·90   HALF PRICE

£54·75   SAVE $\frac{1}{3}$

The difference in price between the most expensive and least expensive pair of trainers is £☐.

3 **a)** Aki scored $\frac{36}{80}$ on a test. In his next test, there were more questions and Aki scored 40%. Did he improve?

I wonder how I can convert eightieths into a percentage. Perhaps I can do it if I simplify $\frac{36}{80}$ first.

b) Discuss how to compare these values.

$\frac{60}{90}$ ◯ 40%

$\frac{5}{11}$ ◯ 50%

0·251 ◯ $\frac{6}{8}$

$\frac{38}{51}$ ◯ 78%

$\frac{5}{10}$ is equivalent to 50%, so I wonder whether $\frac{5}{11}$ is greater than or less than 50%.

I will use a diagram to find the answer.

79

Mixed problem solving

Discover

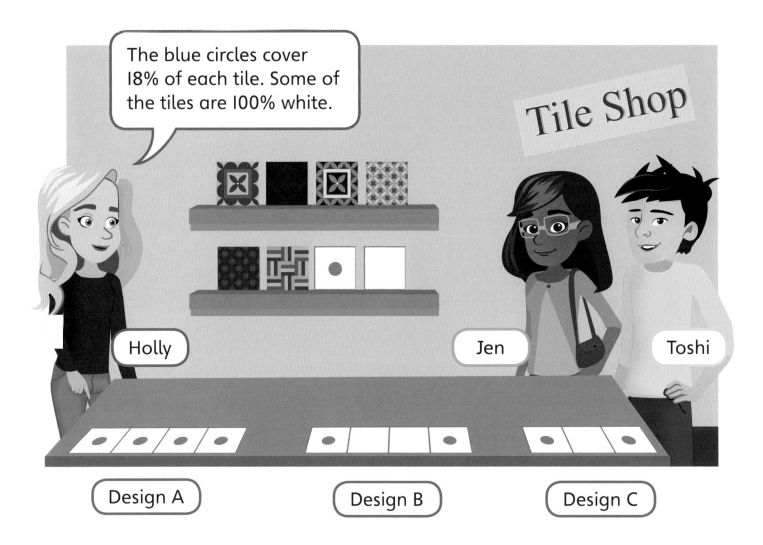

a) What percentage of Design A is blue?

b) Find the percentages of Design B and Design C that are blue.

Share

a) 18 parts out of every 100 are the blue circle.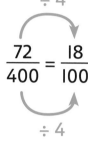

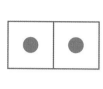 $\div 2$ $\dfrac{36}{200} = \dfrac{18}{100}$ $\div 2$

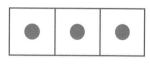 $\div 3$ $\dfrac{54}{300} = \dfrac{18}{100}$ $\div 3$

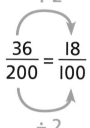 $\div 4$ $\dfrac{72}{400} = \dfrac{18}{100}$ $\div 4$

18% is circles.

18% of each tile in Design A is a blue circle.

18% of Design A is blue.

b)

Design B

    $\div 4$ $\dfrac{36}{400} = \dfrac{9}{100}$ $\div 4$

> I think that to find the percentage I need to make sure that the denominator is 100.

Design C

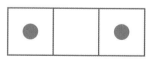 $\div 3$ $\dfrac{36}{300} = \dfrac{12}{100}$ $\div 3$

9% of Design B is blue.

12% of Design C is blue.

Think together

1   This is 15% of the whole shape. How many triangles are in the whole shape?

a) ☐ triangles = 15%

So, ☐ triangles = 5%

5% × ☐ = 100%

So, ☐ triangles × ☐ = ☐ triangles

☐ triangles = 100%

b) What fraction of the whole shape is this?

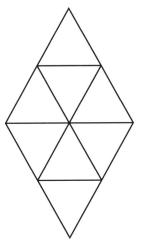

2   Toshi and Amal each have £40 left. How much did they each have to begin with?

I spent $\frac{3}{4}$ of my money.

Toshi

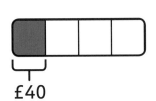

£40

I spent 90% of my money.

Amal

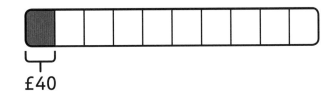

£40

3 **a)** Danny has some clay.

He gives 40% to Bella. He then gives half of what remains to Isla.

Now Danny has 1,200 g of clay.

In grams, how much clay does Bella have?

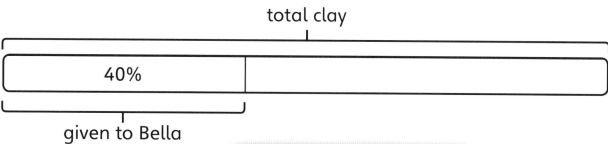

total clay

40%

given to Bella

I think he gives 50% of the total clay to Isla. Am I right?

b) Danny and Bella share out some more clay.

Bella has 45% of the clay and Danny has 1,200 g more than her.

In grams, how much clay is there in total?

I wonder if drawing a comparison bar model would help.

→ **Practice book 6B p59**

End of unit check

1 What percentage of the grid is shaded?

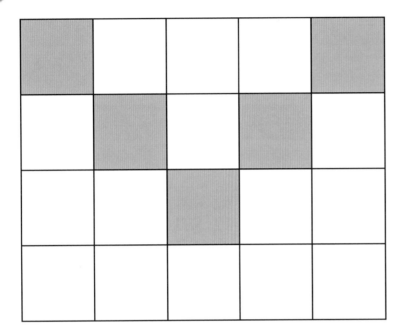

| A | 5% | B | 20% | C | 25% | D | 10% |

2 What is 20% of 45?

| A | 4·5 | B | 9 | C | 900 | D | 20 |

3 What is the missing number?

25% of ☐ = 120

| A | 90 | B | 240 | C | 480 | D | 30 |

4 Which value is not in the shaded part of the number line?

0% 20% 40% 60% 80% 100%

A $\frac{3}{4}$ **B** 0·7   **C** 0·079   **D** $\frac{603}{1,000}$

5 What number did Emma start with?

I am thinking of a number. I halve it. I then find 10%. I finish on 2.

Emma

A 20   **B** 40   **C** 0·1   **D** 40%

6 25,000 fans buy tickets to watch their favourite band in concert. There are three shows.

$\frac{1}{4}$ watch the Thursday show.

30% watch the Friday show.

How many watch the Saturday show?

→ Practice book 6B p62

Unit 9
Algebra

In this unit we will …
- ⚡ Find and write algebraic rules
- ⚡ Write algebraic expressions
- ⚡ Write algebraic formulae
- ⚡ Write and solve algebraic equations
- ⚡ Solve equations that have lots of solutions

Do you remember what this model is called? We will use it to represent different equations. Can you predict what equation is being represented here?

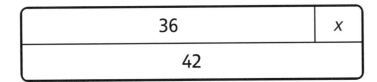

36	x
42	

We will need some maths words. Can you identify and explain the words you already recognise?

sequence   rule   term   algebra

expression   calculation

formula   substitute   generalise

operation   calculate   equation

inverse   solution

We will need to work systematically to find all the solutions to one equation. We can use a table to help us order and record our solutions.

Perimeter of rectangle	If $a =$	Then $b =$
20	$a = 1$	$20 \div 2 - 1 = 9$
20	$a = 2$	$20 \div 2 - 2 = 8$
20	$a = 3$	$20 \div 2 - 3 = 7$

Finding a rule ❶

Discover

> I frog 4 legs; 2 frogs 8 legs; 3 frogs 12 legs; ...

> This nursery rhyme can go on forever. What if we sang 100 verses or more?

❶ **a)** Describe the **rule** for this nursery rhyme. Can it work out the number of legs for any number of frogs?

b) What are the rules for the number of eyes or mouths a frog has?

Share

a)

> The rule is to add on 4 each time: 4, 8, 12, 16, ... but that will take a long time if there are 100 frogs or more!

> This reminds me of missing number calculations like ☐ × 4. I will use a table to find the rule.

Number of frogs	1	2	3	. . .	10	100	1,257	a
Number of legs	1 × 4	2 × 4	3 × 4	. . .	10 × 4	100 × 4	1,257 × 4	a × 4

The number of legs is four times the number of frogs.

Use the letter a to represent the number of frogs.

The rule for the nursery rhyme is: For a frogs, there are a × 4 legs.

a could be 1 frog, 7 frogs, 100 frogs or even 1 million frogs!

> We can use letters to represent a value we do not know for certain, or that can change.

b) Use a to represent the number of frogs for other rules, too.

If there are a frogs, then there are a × 2 eyes and a × 1 mouths.

Think together

1 A bracelet needs 6 sea shells. How many sea shells are needed for 2 bracelets? 3 bracelets? *m* bracelets?

Number of bracelets	1	2	3	. . .	51	*m*
Number of shells needed	1 × ⬜			. . .		

For *m* bracelets, you need ⬜ × ⬜ shells.

2 One bracelet needs a length of 20 cm of red string. What about 2 bracelets? 3 bracelets? *n* bracelets?

Number of bracelets	1	2	3	*n*
Length of string	1 × ⬜			

For *n* bracelets, you need ⬜ × ⬜ cm of red string.

CHALLENGE

3 **a)** Jen is 26 years older than Lee. If *n* is Lee's age, how can Jen's age be represented?

Jen's age					
Lee's age	8	9	10	25	*n*

b) Mr Jones is 47 years older than Ebo. If *y* is Mr Jones's age, how can Ebo's age be represented?

I will decide if I need to add or subtract.

Mr Jones's age	47	57	68	80	*y*
Ebo's age					

c) What rules connect these ages?

Amal	17	27	39	75	?
Mrs Dean	27	37	49	?	100

I could use *p* to represent Amal's age. I think there could be two different rules depending on whose age I know.

91

→ Practice book 6B p64

Finding a rule ❷

Discover

There are 7 geese on the lake. More geese arrive in pairs.

There will be 100 geese if more keep arriving.

Kate

Richard

1 **a)** More pairs of geese land. What is the rule for the number of geese on the lake?

b) Is Richard correct?

Share

a)

$7 + 2 = 9$ $7 + 1 \times 2 = 9$

$7 + 2 + 2 = 11$ $7 + 2 \times 2 = 11$

$7 + 2 + 2 + 2 = 13$ $7 + 3 \times 2 = 13$

$7 + 2 + 2 + 2 + 2 = 15$ $7 + 4 \times 2 = 15$

Geese at the start	7	7	7	7	7
More pairs fly in	1	2	3	100	n
Total geese	$7 + 1 \times 2$	$7 + 2 \times 2$	$7 + 3 \times 2$	$7 + 100 \times 2$	$7 + n \times 2$

If n more pairs land on the lake, there will be $7 + n \times 2$ geese in total.

> The letter n is used here as the letter x would look like the multiplication sign. I wonder if that is confusing in algebra.

> You can write $2n$ instead of $2 \times n$ or $n \times 2$.
>
> This rule can be written $7 + 2n$.

b) No, Richard is not correct because there will never be exactly 100 geese. The rule adds on a multiple of 2 to 7, so the answer will always be an odd number.

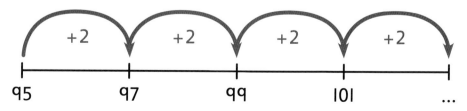

Think together

1 Luis is selling lemonade. Lexi wants to buy some cups of lemonade. How much money is left after Lexi buys one cup of lemonade?

What if she buys 2 cups? 3 cups? a cups?

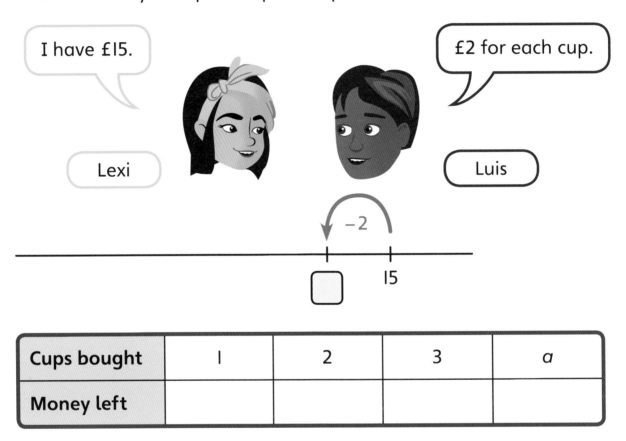

I have £15.

Lexi

£2 for each cup.

Luis

Cups bought	1	2	3	a
Money left				

2 Luis sold 3 cups of lemonade. How much did he earn? What if he sold 4 cups? 5 cups? n cups?

It cost £5 to make all the lemonade.

Luis

If Luis spent £5, I need to subtract that from the amount he earned selling the lemonade.

3 Use sticks to make a square. 1 square needs 4 sticks.
How many sticks are needed for 2 squares? 3 squares?
4 squares? *n* squares?

To make 1 square, 4 sticks are used.

To make 2 squares, 4 + 3 × 1 sticks are used.

To make 3 squares, 4 + 3 × ☐ sticks are used.

To make 4 squares, 4 + 3 × ☐ sticks are used.

To make *n* squares, 4 + 3 × ☐ sticks are used.

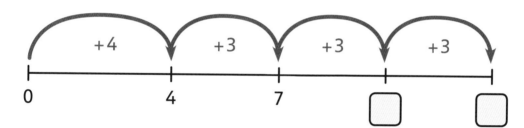

+4   +3   +3   +3

0   4   7   ☐   ☐

Make up your own shapes and come up
with a rule about how many sticks are used.

I remember that 3 × ☐ is the same as ☐ × 3.

95

Using a rule ❶

Discover

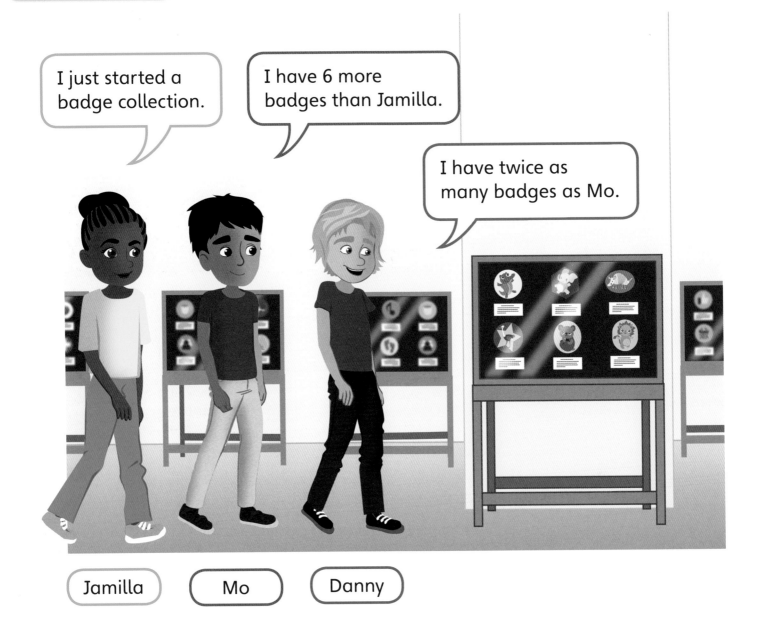

❶ a) If Jamilla has *x* badges, how can the number of badges that Mo and Danny have be represented?

b) How many badges do Mo and Danny have if Jamilla has 4 badges?

Share

a) Jamilla has *x* number of badges. Mo has 6 more badges than Jamilla. We can write that Mo has *x* + 6 badges.

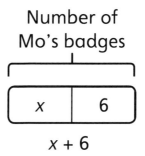

Number of Mo's badges

x	6

x + 6

> Danny has twice as many badges so I think we can write that he has 2*x*.

> I do not think that is right. Danny has twice as many badges as Mo, **not** Jamilla, so it is not 2*x*. I will show it with a bar model.

Number of Danny's badges

x	6	*x*	6

(*x* + 6) × 2

We can write that Danny has (*x* + 6) × 2 badges.

b) Jamilla has 4 badges, so *x* = 4.

Number of Mo's badges

4	6

4 + 6 = 10

Number of Danny's badges

4	6	4	6

(4 + 6) × 2 = 20

If Jamilla has 4 badges, Mo has 10 badges and Danny has 20 badges.

Think together

1 Amelia and Bella collect snow globes. Bella has 3 times as many snow globes as Amelia, plus 2 more.

a) Complete the rule.

If Amelia has *x*, then Bella has ☐◯☐.

Amelia: ☐

Bella: | ☐ | ☐ | ☐ | 2 |

b) How many snow globes does Bella have if Amelia has 29?

If Amelia has 29 snow globes, then Bella has ☐.

2 Reena inputs 4 into this function machine. First it adds 10, then the result is halved.

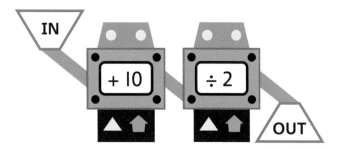

Complete the table.

Input	4	10	40	41	*x*
Output	7				

Write the rule for the output for *x*.

(*x* ◯☐) ◯☐

98

3 Lexi is investigating this function machine.

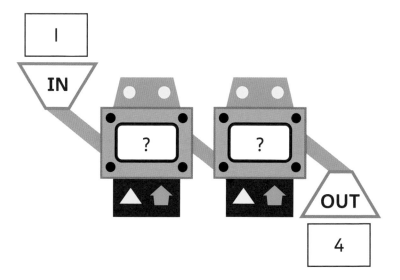

What could the two functions be?

Complete the table for different inputs, based on your functions.

Input	1	2	3	10	x
Output	4				

I wonder if there are different functions that will work.

I will try the operations in a different order.

→ **Practice book 6B p70**

Using a rule ②

Discover

① **a)** Complete the table.

Write the rule for *n* stars.

How many points will you have for Level I if the value of *n* is 13?

b) What happens to the score if the value of *n* increases by 10 to 23?

Number of ⭐	Points for Level I
I	
2	
3	
4	
n	

Share

When 5 × *n* is written as 5*n*, it is called an **expression**.

When a specific value is given for *n*, you **substitute** the value for *n* into the rule. So here you substitute 13 for *n*.

a)

Number of 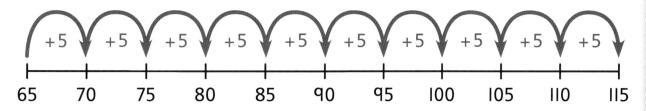	Points for Level I
I	5 × I = 5
2	5 × 2 = 10
3	5 × 3 = 15
4	5 × 4 = 20
n	5*n*

If the value of *n* is 13, that means 13 stars have been collected.

The rule is 5*n* which means 5 × *n* or *n* × 5.

When *n* = 13, 5*n* = 5 × 13.

5 × 13 = 65

If the value of *n* is 13, you will have 65 points.

b) Now *n* = 23.

Method I

Work out 5 × 23.

5 × 20 = 100

5 × 3 = 15

If *n* = 23, you will have 115 points.

Method 2

If *n* increases by 10, that means 10 more stars have been collected. So the number of points will increase by 10 × 5.

If *n* = 23, the score will be 65 + 50 = 115 points.

+5 +5 +5 +5 +5 +5 +5 +5 +5 +5

65 70 75 80 85 90 95 100 105 110 115

Think together

1 **a)** On Level 2, each lightning bolt gives 15 points. Write the rule for *m* lightning bolts collected.

b) Substitute *m* = 9, *m* = 10, and *m* = 11 into the rule.

When *m* = 9, points = ☐.

When *m* = 10, points = ☐.

When *m* = 11, points = ☐.

Number of lightning bolts collected	Points for Level 2
1	
2	
3	
m	

2 To reach Level 3, you need 100 points.

a) In Level 3, if you hit a spike, you lose 3 points. What is the rule for the score after hitting *k* spikes?

Spikes hit	Points remaining
1	
2	
3	
k	

> I need to start with 100 points. Then each spike hit takes away some points from the score.

b) Substitute *k* = 10, *k* = 20, and *k* = 30 into the rule.

c) What happens when you substitute 33 and 34 into the rule?

3 **a)** Complete the table. What patterns do you notice?

	2x + 1	2x − 1
Substitute x = 1		
Substitute x = 15		
Substitute x = 101		
Substitute x = 1,213		

2x + 1 makes a lot of odd numbers for the values I substituted. I wonder if that always happens.

I think I can explain by thinking about the 2x part of the rule.

b) Substitute different values of x into the rules 10x + 1 and 10x − 1. What do you notice?

→ **Practice book 6B p73**

Using a rule 3

Discover

If this tap leaks for I hour, you will lose 20 ml of water.

Amal

Sofia

1 **a)** How much water will this tap lose in one day?

b) There is 50 ml water in the jug. The tap drips for *n* more hours. Write a rule for how much water there will be in the jug after *n* more hours.

Share

a) If t is the number of hours, the rule is $20t$.

> I do not need the whole table.
>
> I can substitute the value I need into the expression $20t$.

Number of hours	Water lost (ml)
1	$20 \times 1 = 20$
2	$20 \times 2 = 40$
3	$20 \times 3 = 60$
4	$20 \times 4 = 80$
. . .	. . .

There are 24 hours in a day, so $t = 24$.

$20t = 20 \times 24 = 480$ ml

The tap will lose 480 ml of water in one day.

b)

> I think this is $50 + 20$ for every hour. So that is $70n$.

> I do not think that is quite right. I will draw a diagram to show this.

Number of hours	Water in the measuring jug (ml)
1	$50 + 20 \times 1 = 70$
2	$50 + 20 \times 2 = 90$
3	$50 + 20 \times 3 = 110$
4	$50 + 20 \times 4 = 130$
. . .	. . .

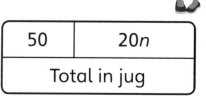

The rule is $50 + 20n$.

Think together

1 A water barrel has 150 litres of water. It loses 2 litres per day. How much is left after a week? How about after *x* days?

I will say *x* is the number of days.

The expression must be something like 150 − ☐ *x*.

Number of days	Litres of water left in the barrel
1	
2	
3	
. . .	. . .
7	
x	

2 An hourglass contains 720 grams of sand. 20 grams of sand trickles per minute.

a) Complete the expression to show the amount of sand left after *y* minutes.

720 ◯ ☐ *y*

720 g

?	Sand trickled down

b) Use a table to show how many grams of sand are left.

Time taken	Your calculation	Sand left in the hourglass (grams)
10 minutes		
20 minutes		
0·5 hours		

CHALLENGE

3 **a)** Copy each bar model. Write an expression in each ▢.

$\square$	
m	10

m	
$\square$	30

m		
$\square$	$\square$	$\square$

$\square$		
m	m	12

> I think each ▢ will have an expression involving m.

> I wonder if they all use the same operation.

b) Substitute $m = 48$ into each expression.

Now try substituting $m = 24$ into each expression. What happens?

→ **Practice book 6B p76**

Formulae

Discover

Shapes are everywhere if you look. This football is composed of 12 black pentagons and 20 white hexagons. The structure is exactly the same as this carbon molecule!

Mrs Dean

I wonder if the same rule can be used to find the perimeter of the shapes on a football and the shapes on a carbon molecule.

1 **a)** Write the rule to work out the perimeter of a regular pentagon and a regular hexagon. Use a to represent the length of each side.

b) Substitute the value 6 cm for the side length of each shape.

Now try 12 cm. What do you notice?

Share

a) The perimeter of a polygon is the sum of all the sides.

For a regular polygon, all sides are equal.

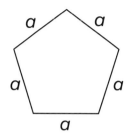 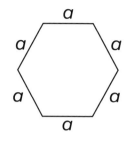

> The two shapes have sides the same length.
>
> I will use a to represent the length of the sides.

Perimeter of a regular pentagon = $a + a + a + a + a = a \times 5 = 5a$

Perimeter of a regular hexagon = $a + a + a + a + a + a = a \times 6 = 6a$

This kind of rule is called a **formula**. It gives you a way to work out the whole perimeter, if you know the value of a.

b)

Shape	6 cm sides	12 cm sides
(pentagon)	Perimeter $= 5a$ $a = 6$ $5a = 5 \times 6 = 30$ cm	Perimeter $= 5a$ $a = 12$ $5a = 5 \times 12 = 60$ cm
(hexagon)	Perimeter $= 6a$ $a = 6$ $6a = 6 \times 6 = 36$ cm	Perimeter $= 6a$ $a = 12$ $6a = 6 \times 12 = 72$ cm

When the value of a is doubled, the perimeter of each shape is also doubled.

Think together

1 Jen is building a rectangular fence for her garden.

a) Find the formula for the rectangular perimeter.

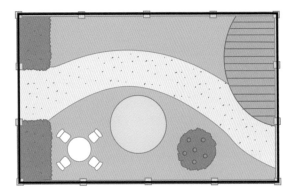 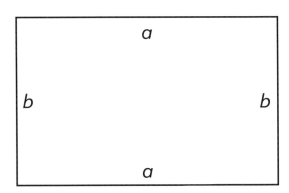

The formula for the perimeter of a rectangle is

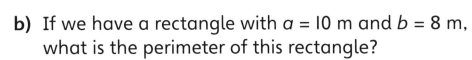

I can write the formula in two different ways.

b) If we have a rectangle with $a = 10$ m and $b = 8$ m, what is the perimeter of this rectangle?

Perimeter of the rectangle = ☐ m

2 Now write a formula for the area of Jen's rectangular garden.

Area of the rectangle = ☐

Substitute the values $a = 7$ m, $b = 7$ m into the formula.

If $a = 7$ m and $b = 7$ m:

Area of the rectangle = ☐ m²

CHALLENGE

3 Isla is investigating some equivalent calculations.

a) Explain the pattern and use algebra to represent the ideas.

$$1 + 2 = 2 + 1$$
$$10 + 8 = 8 + 10$$
$$15 + 20 = 20 + 15$$

I will use letters to represent the values in the calculations.

In the first calculation, the number 2 appears on both sides of the equation. I will use the same letter to represent this.

Test your ideas by substituting different values into your expressions.

b) Use algebra to describe this pattern.

$$(8 + 6) \times 5 = 8 \times 5 + 6 \times 5$$
$$(10 + 6) \times 8 = 10 \times 8 + 6 \times 8$$
$$4 \times (10 + 2) = 4 \times 10 + 4 \times 2$$

111

Solving equations ❶

Discover

How much did the kayak cost?

Surf's Up

1 surfboard £ 80

1 kayak

Total: £230

Toshi

❶ **a)** Toshi writes 80 + x = 230 to work out the cost of the kayak.

What does x represent?

What does 80 + x represent?

What does 80 + x = 230 mean?

b) Show different methods to find the cost of the kayak.

Share

a)

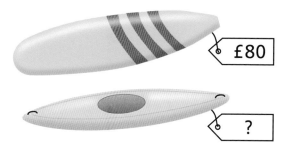

£80

?

The *x* represents the unknown cost of the kayak.

80 + *x* represents the total cost of the surfboard and kayak.

Total: **£230**

80 + *x* = 230 means that the total of both items is equal to £230.

80 + *x* = 230 is called an **equation**, because 80 + *x* **equals** 230.

b) Finding the unknown value is called solving the equation.

I will substitute different values for *x* until I find one where 80 + *x* = 230.

If *x* is:	80 + *x* will be:
100	180
110	190
120	200
130	210
140	220
150	230

This is just a missing number problem. 80 + ☐ = 230. I can represent this using a bar model.

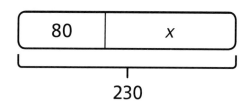

80	*x*

230

230 − 80 = 150

The cost of the kayak is £150.

Think together

1 Amelia bought a dinghy. She spent £85. The price had been reduced by £35.

Represent the **usual** cost of the dinghy as y.

Complete and solve the equation.

$y - \boxed{} = \boxed{}$

$y = \boxed{}$

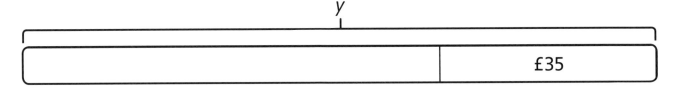

The dinghy usually costs £$\boxed{}$.

2 The shop has 3 crates of wetsuits. Each crate contains an equal number. In total, there are 360 wetsuits.

a) Use s to represent the number of wetsuits in 1 crate.

Write an equation.

$\boxed{} \bigcirc s = \boxed{}$

b) Work out how many wetsuits are in 1 crate.

$\boxed{} \bigcirc \boxed{} = \boxed{}$

If s is:	3s will be:
80	240
90	270
100	
110	
...	...

3 These children all thought of a mystery number. Then ...

I added 9 to mine.

Kate

I found the number that was 9 less than mine.

Alex

I worked out the number 9 times bigger than mine.

Richard

I divided mine by 9.

Aki

They all ended up with 45.

Match each equation with the correct person, and find their mystery numbers.

$45 = 9h$ $9 + i = 45$ $45 = k \div 9$ $j - 9 = 45$

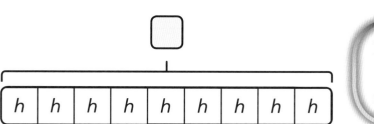

I will try drawing a bar model to help solve each equation.

→ Practice book 6B p82

Solving equations ②

Discover

1 a) What is the mystery weight?

Explain how knowing this helps you solve the equation $x + 36 = 42$.

b) Solve the equation $36 + x = 42$.

Share

a)

I drew the scales and then crossed out one weight from each side at the same time. It is like solving an equation.

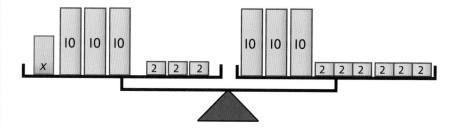

$x + 36 = 42$

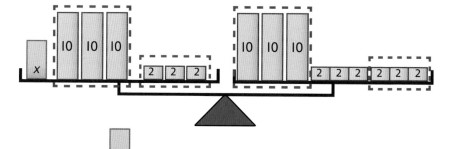

Subtract 36 from each scale.

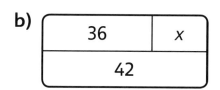

$x = 6$

The mystery weight is 6 kg.

Finding the mystery weight means you can replace x with 6 to solve the equation.

I can check my answer.
$6 + 36 = 42$. Correct!

b)

36	x
42	

$36 + x = 42$

I drew a bar model. It shows me that 36 and the unknown (x) added together make 42.

To find the missing number, subtract 36 from 42.

$x = 42 - 36$, so $x = 6$.

Think together

1 **a)** Remove weights from both sides of the balance to solve the equation

$240 = y + 48$.

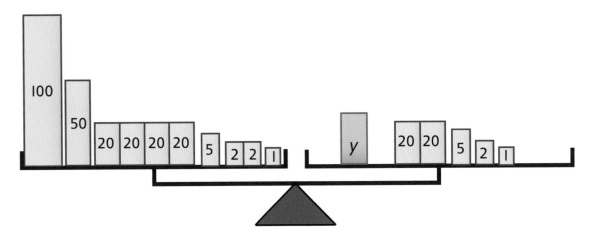

$240 = y + 48$

Subtract ☐ from each scale.

$y =$ ☐

b) What equation is represented here?

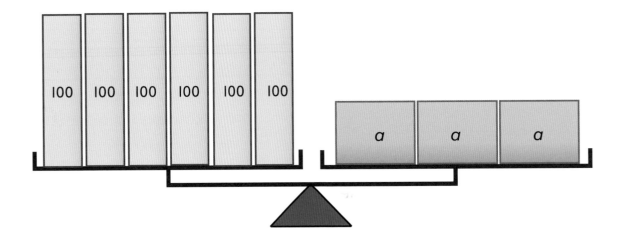

$600 =$ ☐

2 Use the bar models to solve the equations.

a) $t + 6 = 24$

$t =$ ☐

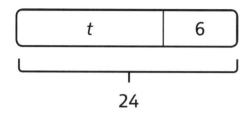

| t | 6 |

24

b) $24 = 6m$

$m =$ ☐

| m | m | m | m | m | m |

24

3 **a)** Max wants to solve the equation $x - 10 = 36$.

Which bar model helps you to solve Max's equation?

Explain your reasoning.

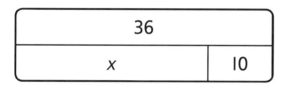

| 36 | |
| x | 10 |

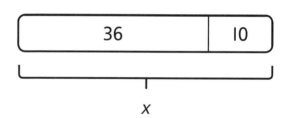

| 36 | 10 |

x

What is the solution to Max's equation?

b) Solve the following equations.

$y + 10 = 25$ $k - 10 = 25$

$10 + g = 25$ $25 - h = 10$

$5a = 30$ $4b = 600$

For each of these equations I am going to try to draw a bar model to help me work out the unknown value.

→ **Practice book 6B p85**

Solving equations ③

Discover

I a) Represent Kate's number as *a*.

Write down an equation that you can use to find Kate's number.

b) What was Kate's number?

Share

a) Start with Kate's number. a

Double a, which is the same as multiplying by 2. $2a$

Now add 5. $2a + 5$

The answer is 19, which can be written $2a + 5 = 19$.

This is the equation that needs to be solved.

b) $2a + 5 = 19$

$\quad\quad -5 \quad\ -5$

$\quad\quad 2a = 14$

So a must be 7, because $2 \times 7 = 14$.

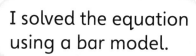

I used the
balance model.

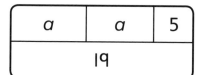
$2a + 5 = 19$

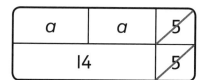
$-5 \quad\quad -5$

I solved the equation
using a bar model.

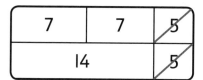
$2a = 14$

$\div 2 \quad \div 2$

Kate's number was 7.

121

Think together

1 What is Lexi's number?

Lee: Multiply your number by 3 and add 4.

Lexi: My answer is 19.

3x + ☐ = ☐

   − ☐   − ☐

3x = ☐

 ÷ ☐   ÷ ☐

x = ☐

x	x	x	4
19			

2 Solve the following equations.

a) 5m + 3 = 58

b) 17 = 2q + 5

m	m	m	m	m	3
58					

3 **a)** A kettle has a capacity of 600 ml.

Richard fills 3 cups from the kettle.

He has 30 ml left over.

What is the capacity of 1 cup?

Use a for the capacity of the cup.

$$\boxed{}\,a\,\bigcirc\,\boxed{} = 600 \text{ ml}$$

$$3a = \boxed{}$$

$$a = \boxed{}$$

600 ml			
a	a	a	

30 ml

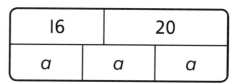

b) Solve the following equations.

i) $3a - 20 = 16$

ii) $20 = 2p + 8$

iii) $4c - 6 = 14$

iv) $2p + 8 = 21$

16		20
a	a	a

I will draw different bar models to represent the equations.

123

Solving equations ④

Discover

Holly

> We need 20 metres of fencing to make a rectangular alpaca enclosure.

① **a)** The perimeter of the enclosure is 20 m.

Write an equation for the perimeter of the enclosure. Find different solutions for a and b.

b) Which solution has the greatest area?

Share

a)

The formula is
perimeter = 2a + 2b or
perimeter = (a + b) × 2.

So a + b must be equal to 10.

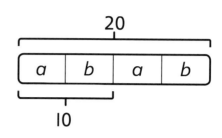

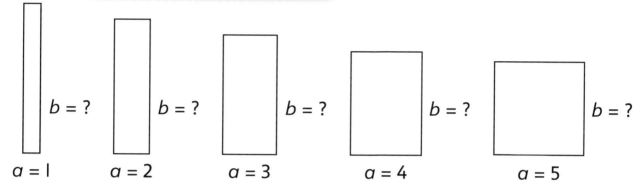

b = ?   b = ?   b = ?   b = ?   b = ?

a = 1   a = 2   a = 3   a = 4   a = 5

Perimeter of rectangle	a = ?	b = ?
20	1	9
20	2	8
20	3	7
20	4	6
20	5	5
20	6	4

b) Area is a × b.

The greatest area for this enclosure
is 5 × 5 = 25 m². That is a square
enclosure.

I will think in order. If a = 1,
then b = 9. I will continue until
I start to repeat numbers.

Think together

1 One alpaca eats 15 kg of hay and grass every day. *m* represents the weight of hay, *n* represents the weight of grass. How many kg could there be for *m* and *n*?

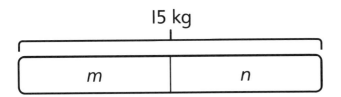

15 kg

m   *n*

m + ☐ = ☐

m = ?	*n* = ?
0	15 − 0 = 15
1	15 − 1 = 14

2 The area of this rectangle is 36 m².

Write the equation for the area and find all the solutions.

y

z

☐ ◯ ☐ = ☐

y = ?	*z* = ?

3 **a)** Max is finding solutions to $x + y = 10$.

He found $x = 3$, $y = 7$ as a solution, and marked this on the grid.

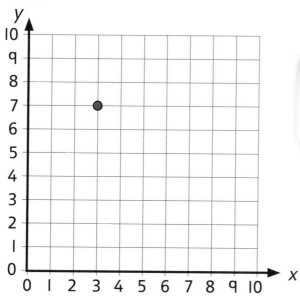

I will draw a table to find all the possible numbers for x and y, starting with $y = 10$.

Find other solutions.

What do you notice?

b) Investigate solutions to this equation, using all four quadrants.

$y = x + 1$

→ **Practice book 6B p91**

Solving equations ⑤

Discover

The total number of legs of the chickens and rabbits in this shelter is 10.

Amelia

Jamilla

I **a)** How many chickens and how many rabbits are there in the shelter?

Say there are *x* chickens and *y* rabbits.

Write an equation and find all solutions.

b) Explain why there cannot be 2 or 4 chickens in the shelter.

Share

a) Legs of x chickens + legs of y rabbits = 10

$2x + 4y = 10$

Number of chicken legs
1 chicken, 2 legs
2 chickens, 4 legs
3 chickens, 6 legs
4 chickens, 8 legs
5 chickens, 10 legs
x chickens, $2x$ legs

Number of rabbit legs
1 rabbit, 4 legs
2 rabbits, 8 legs
3 rabbits, 12 legs
4 rabbits, 16 legs
5 rabbits, 20 legs
y rabbits, $4y$ legs

The possible solutions are: 1 chicken and 2 rabbits or 3 chickens and 1 rabbit.

Could there be 5 chickens and 0 rabbits?

b) Solve the equation $2x + 4y = 10$ with different values for x to see different possible solutions.

Number of legs in total	If x = ?	Then y = ?
10	1	2
10	2	1.5
10	3	1
10	4	0.5
10	5	0

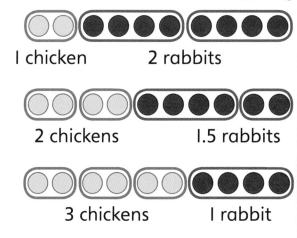

There cannot be 2 or 4 chickens in the shelter, because this would mean there would have to be half a rabbit.

Think together

1 There are two kinds of table. The class needs 30 seats. Find different solutions.

There are *m* tables for 6 people and *n* tables for 4 people.

◻ *m* + ◻ *n* = 30

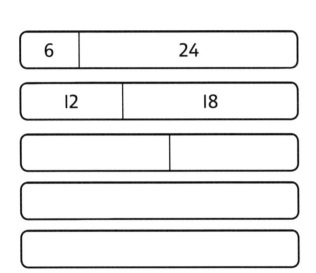

6	24

12	18

Total number of people	If *m* = ?	Then *n* = ?
30	1	
30	2	

2 There are some chickens and rabbits in a field. There are 35 heads and 94 legs in total. How many chickens and how many rabbits are in the field?

35 heads in total means the total number of chickens and rabbits is 35.

Number of chickens	Number of rabbits	Number of legs
1	34	(1 × 2) + (34 × 4) = 138
2	33	(2 × 2) + (33 × 4) = 136

3 **a)** Find five different solutions to each equation.

$100 = 20a + 10b$ $\qquad\qquad$ $100 = 10c - 20d$

Can you find all the possible solutions?

I noticed a pattern in how a and b are related.

I wonder how c and d are related.

b) Describe the patterns in the solutions to these equations.

$x + 30 = y - 70$

$20s = 100 - 2t$

I will start by choosing a value for x, and then see if I can work out what y must be.

131

End of unit check

 Which expression represents the number of wheels on *n* tricycles?

A $n + 3$ **B** $3 - n$ **C** $3n$ **D** $n \times n \times n$

 What is the output when *a* is the input?

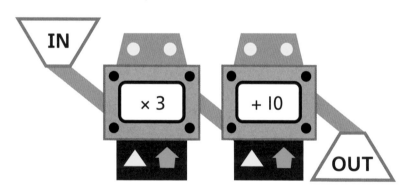

A $a + 10 \times 3$ **B** $3a$ **C** $a + 10$ **D** $3a + 10$

3 What is the result when you substitute $x = 8.5$ into the expression $120 - 10x$?

A 35 **B** 85 **C** 111.5 **D** 110

4 What equation is represented by this bar model?

f	f	f	21
56			

A $3f + 56 = 21$ B $3f - 56 = 21$ C $56 = 3f$ D $21 + 3f = 56$

5 Solve the equation $100 = 25s - 25$.

A $s = 5$ B $s = 0$ C $s = 125$ D $s = 75$

6 Each bar model is of equal length. What is the value of w?

w	18	w	9	w

45	w	18

→ **Practice book 6B p97**

Unit 10
Measure – imperial and metric measures

In this unit we will …

⚡ Choose the most appropriate metric units of measurement to measure different things

⚡ Convert between metric units, between imperial units and from one to the other

⚡ Solve problems involving metric units

⚡ Recognise the difference between metric and imperial units of measurement and what they are worth

What is 1 inch about the same as?
What are 5 inches about the same as?

5 inches

1 inch	1 inch	1 inch	1 inch	1 inch
2·5 cm	2·5 cm	2·5 cm	2·5 cm	2·5 cm

Here are some maths words we will be using. Which words are new to you?

metric   imperial

units of measurement (or measure)

grams (g)   kilograms (kg)   pounds (lbs)

ounces (oz)   mass   millilitres (ml)

litres (l)   pints   capacity   millimetres (mm)

centimetres (cm)   metres (m)   kilometres (km)

inches (in)   feet (ft)   yards   miles   length

convert   conversion table   conversion graph

If there are 100 cm in a metre, how would you convert 4·5 metres into centimetres?

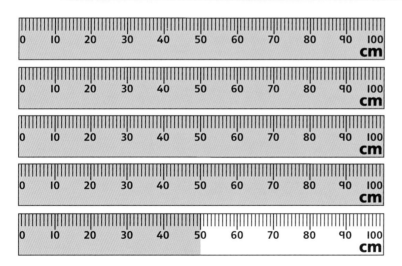

1 m = 100 cm

Metric measures

Discover

1 **a)** Look at the weighing scales and the measuring jug.

   What units of measure do you think they show?

   b) Estimate the length of the oven tray.

Share

a) It is always important to use the correct unit of measurement. For different situations, some units of measure are more appropriate than others.

Mass is measured in grams or kilograms.

I know that 1 kg is about the mass of a bag of sugar, so 150 kg would be far too heavy.

The sugar must weigh 150 grams.

Capacity is measured in millilitres or litres.

I know that 1 ml is about the capacity of a drop of water, so 0·75 ml would be far too small.

The jug must contain 0·75 litres.

The units of measure for the weighing scales and the measuring jug are grams and litres.

b) The length of a school ruler is 30 cm. The oven tray is slightly longer than this.

Estimation is about using the facts that you know to help estimate what you do not know.

So you can estimate that the tray is about 35 cm long.

Think together

1 Estimate the number of millilitres the glass can hold.

There is a little water left in the bottle!

The bottle holds ▢ ml.

The glass holds slightly less than the bottle.

The glass holds about ▢ ml.

2 What unit of measurement should be used for each of these items?

| ml | g | kg | g | l |

138

3 **a)** Which metric unit of measurement do you think is most appropriate to use to measure the length of your classroom?

Share your thoughts with your group.

Did the rest of your group have the same ideas as you?

b) What if you had to measure the length of your shoe or the distance from your house to school? Would you use the same unit of measure or a different unit?

Explain your answer.

c)

I have measured this piece of spaghetti in centimetres. It is 26 cm long.

Jamilla

I prefer to use metres. The spaghetti is 0·26 m long.

Aki

Both measurements are equal, but they use different metric units.

Who do you think has used the more appropriate unit of measurement?

Explain why.

139

→ Practice book 6B p100

Converting metric measures

Discover

This piece of wood is between 1·2 m and 1·3 m long.

Toshi

1,000 g = 1 kg
1,000 ml = 1 l
100 cm = 1 m

CEMENT
40,500 g

9·25 l

1 **a)** How could you label the cement and the bucket using different units of measurement?

   b) How many centimetres long could the piece of wood be?

Share

a) To **convert** between units of measurement, you need to know what one unit is worth.

To convert from a larger unit to a smaller unit, you multiply. To convert from a smaller unit to a larger unit, you divide.

grams > kilograms

$40,500 \div 1,000 = 40 \cdot 5$

grams is a smaller unit of measure than kilograms, so divide

$1 \text{ kg} = 1,000 \text{ g}$, so divide by 1,000

litres > millilitres

$9 \cdot 25 \times 1,000 = 9,250$

litres is a larger unit of measure than millilitres, so multiply

$1 \text{ l} = 1,000 \text{ ml}$, so multiply by 1,000

40,500 g can be written as 40·5 kg.   9·25 l can be written as 9,250 ml.

TTh	Th	H	T	O	•	Tth	Hth
4	0	5	0	0	•		
			4	0	•	5	

TTh	Th	H	T	O	•	Tth	Hth
				9	•	2	5
	9	2	5	0	•		

b) metres ⟶ centimetres

larger unit ⟶ smaller unit, so multiply

$1 \cdot 2 \times 100 = 120$ $1 \cdot 3 \times 100 = 130$

The piece of wood is between 120 cm and 130 cm long.

It could be 123 cm long.

Think together

 a) How many litres of paint does the tin hold?

ml ⟶ l

smaller unit ⟶ larger unit, so divide

1 l = ☐ ml, so ÷ by ☐

4,800 ÷ ☐ = ☐

The tin contains ☐ litres of paint.

4,800 ml

Th	H	T	O	•	Tth	Hth
4	8	0	0	•		
			4	•	8	

b) What is the length of the hose in centimetres?

m ⟶ cm

larger unit ⟶ smaller unit, so ◯

1 m = ☐ cm, so ◯ by ☐

2·75 ◯ ☐ = ☐

The hose is ☐ cm long.

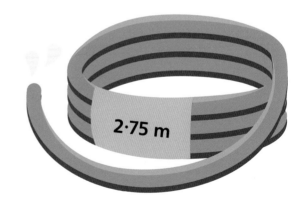

2·75 m

2 Which of these conversions require you to divide by 1,000?

6,000 mm = ☐ cm

500 g = ☐ kg

6·5 l = ☐ ml

3 Olivia and Ebo have been converting different measurements.

I multiplied by 1,000 to convert between my two units!

I needed to divide by 100 for mine!

Olivia

Ebo

Which units of measurement could Olivia and Ebo have converted between?

Explain your answer.

I know I need to multiply when converting to a smaller unit and divide when converting to a larger unit.

I am going to use numbers to work out what units they may have been thinking about.

143

Problem solving – metric measures

Discover

My watering can holds 1·4 l.

My watering can holds 1,250 ml.

Mine holds half as much as yours, Bella!

4 litres of water

Bella   Reena   Richard

1 **a)** How much more water does Bella's watering can hold than Reena's?

b) If the children use water from the container to fill their watering cans, how much water will be left in the container?

Share

a) To compare or calculate between measurements, first convert them into the same units.

litres × 1,000 = millilitres

1·4 l × 1,000 = 1,400 ml

Bella's watering can holds 1,400 ml.

Reena's watering can holds 1,250 ml.

1,400 − 1,250 = 150

Bella's watering can holds 150 ml more water than Reena's.

> I am going to make sure both measurements are in millilitres before I work out the answer.

1,400 ml	
1,250 ml	?

b) We need to find out how much water Richard's watering can holds. His watering can holds half as much water as Bella's does.

1,400 ÷ 2 = 700

Richard's watering can holds 700 ml.

The watering cans hold 3,350 ml in total.

Convert all of the measurements to millilitres to find the answer.

4 l × 1,000 = 4,000 ml

The container holds 4,000 ml.

4,000 − 3,350 = 650

There will be 650 ml of water left in the container.

```
     Th  H  T  O
      1  4  0  0
      1  2  5  0
  +      7  0  0
     _____
      3  3  5  0
         1
```

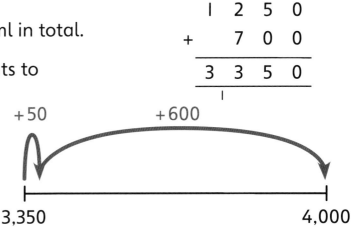

There will be 650 ml of water left in the container.

Think together

1 The water carrier holds 3 litres of water but 600 ml of water has leaked onto the floor.

How many millilitres are left in the water carrier?

3 × 1,000 = ☐ ml

☐ – 600 = ☐

There are ☐ ml of water left.

> The units of measurement are different, so to give the answer in millilitres you need to convert all the units into millilitres.

2 4 kg of compost is divided among 5 trees.

How many grams of compost will each tree get?

4 × 1,000 = ☐

4 kg = ☐ g

☐ g of compost is shared among the trees.

☐ ÷ 5 = ☐

Each tree will get ☐ grams of compost.

3 A garden hose is 8 metres long.

A flower bed is 9·2 metres away from the water tap.

The hose is too short!

How many centimetres longer does the hose need to be?

4 1·4 kg of grass seed is tipped out of a 6 kg bag.

How many grams of grass seed are left?

I am going to convert both amounts into grams before I subtract them.

I am going to subtract the amounts first and then convert the answer into grams at the end.

a) Work out the answer using both methods.

b) Does it matter when you convert the units of measurement?

147

Miles and km

Discover

1 a) What are the distances to Petite Ville and Grande Montagne in miles?

   b) What other facts about miles and kilometres can you work out?

   How could a graph help you to find more facts?

Share

a)

8 km
5 miles

So

16 km

8 km	8 km
5 miles	5 miles

10 miles

16 km ÷ 8 km = 2

2 × 5 miles = 10 miles

Petite Ville is 10 miles away.

40 km

8 km	8 km	8 km	8 km	8 km
5 miles	5 miles	5 miles	5 miles	5 miles

25 miles

40 km ÷ 8 km = 5

5 × 5 miles = 25 miles

Grande Montagne is 25 miles away.

b)

I am going to use the fact that 5 miles is about 8 km to make a **conversion table**. Then I can find lots of new facts.

Miles	km
5	8
10	16
15	24
20	32
25	40
30	48

The values from the table can also be shown on a **conversion graph**.

Here are some new facts shown by the graph:

2·5 miles is about 4 km.

10 km is about 6 miles.

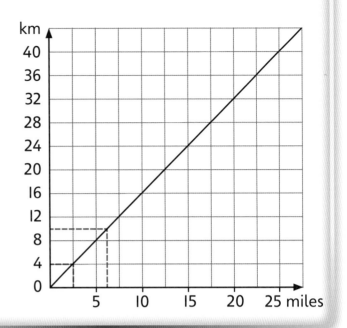

Think together

① How many miles away is the airport?

$56 \div 8 = \boxed{}$

56 km

8 km	8 km	8 km	8 km	8 km	8 km	8 km

5 miles	5 miles

$\boxed{} \times 5 = \boxed{}$

The airport is $\boxed{}$ miles away.

2 A railway station is 65 miles away.

How many km should the sign show?

Railway station ? km

65 miles

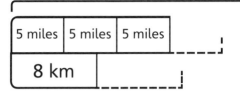

| 5 miles | 5 miles | 5 miles |

8 km

$65 \div 5 = \boxed{}$

$\boxed{} \times 8 = \boxed{}$

The sign should show $\boxed{}$ km.

3 1 mile is about 1·6 km.   1 km is about 0·62 miles.

Bella and Zac are calculating the number of kilometres in 7 miles.

We need to multiply 7 by 0·62.

Bella

That is not right. We need to multiply 7 by 1·6 instead.

Zac

Who is correct?

Explain your answer.

151

Imperial measures

Discover

1 a) 1 inch is about 2·5 cm.

How wide is the pizza in centimetres?

b) 1 pound is about 450 g.

How can Holly weigh 2·5 pounds of sugar with her scales?

Share

a) Inches are **imperial** measures. I inch is about 2·5 cm.

Although metric measures are used most often, imperial measurements are still used in everyday life. You can convert between them.

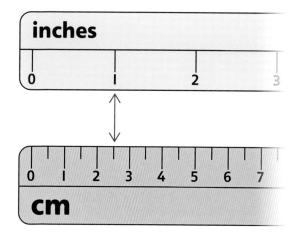

inches

10 inches

I inch	I inch	I inch	I inch	I inch	I inch	I inch	I inch	I inch	I inch
2·5 cm	2·5 cm	2·5 cm	2·5 cm	2·5 cm	2·5 cm	2·5 cm	2·5 cm	2·5 cm	2·5 cm

2·5 cm × 10 = 25 cm   A 10-inch pizza is about 25 cm wide.

b)

I am going to double 450, then halve 450 and add the two answers together.

$2\frac{1}{2}$ lbs

I lb	I lb	$\frac{1}{2}$ lb
450 g	450 g	225 g

450 g × 2 = 900 g   450 g ÷ 2 = 225 g

900 g + 225 g = 1,125 g

Holly should weigh 1,125 grams of sugar.

153

Think together

Remember, a conversion graph helps us convert between two units of measure.

 a) Use the conversion graph to convert 4·5 pounds into kilograms.

4·5 pounds is about the same as ☐ kilograms.

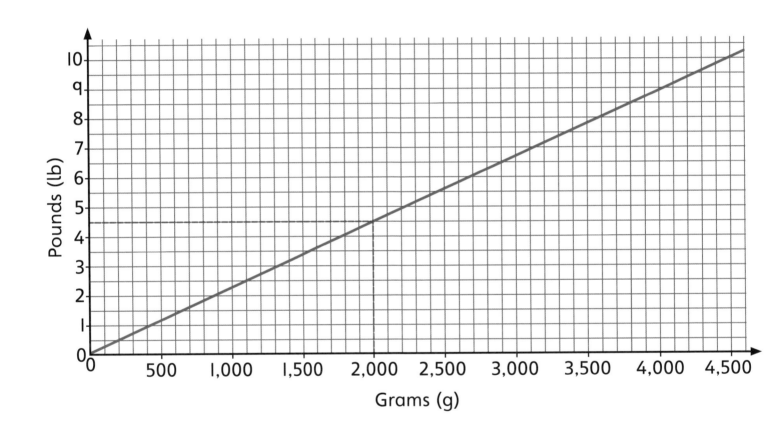

b) Use the graph to convert these amounts.

2 pounds is about the same as ☐ g.

4,500 g is about the same as ☐ pounds.

2·5 kg is about the same as ☐ pounds.

2 How many litres of milk are in a four-pint carton?

550 ml × 4 = ☐ ml

☐ ml ÷ 1,000 = ☐ litres

There are ☐ litres of milk in a four-pint carton.

I pint is about the same as 550 ml.

4 pints

MILK

3 There are 12 inches in 1 foot.

Ebo's dad is 6 feet tall.

How tall is Ebo's dad in centimetres?

'Foot' and 'feet' are the same unit of measurement. 'Feet' is used when there is more than one foot.

I need to use a fact about inches and centimetres to help me find the answer.

155

End of unit check

1 Which of these is the best estimate for the mass of a tomato?

A 1,200 g   **B** 120 ml   **C** 120 kg   **D** 120 g

2 What would you do to convert 8·5 litres into ml?

A Divide by 1,000   **C** Divide by 10

B Multiply by 1,000   **D** Multiply by 10

3 Reena weighs a book.

She adds a second book to the scales. The scales now show 2·4 kg.

How heavy is the second book in grams?

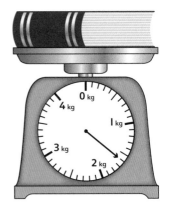

A 4 g   **B** 80 g   **C** 800 g   **D** 0·8 g

4 Andy's grandparents live 30 miles away.

How many kilometres away do they live?

5 miles is about 8 km.

A 40 km **B** 150 km **C** 240 km **D** 48 km

5 Which of these statements is false?

A 1 cm is about 2·5 inches.

B 1 kilogram is about 2·2 pounds.

C 1 mile is about 1·6 km.

D 1 inch is about 2·5 cm.

6 A pet shop owner is weighing out 1·5 kg of dog biscuits.

How many grams of dog biscuits are on the scales already?

☐ grams

How many more grams of dog biscuits are needed to make 1·5 kg?

☐ grams

157

→ Practice book 6B p115

Unit 11
Measure – perimeter, area and volume

In this unit we will ...

⚡ Find and draw shapes with the same area or perimeter

⚡ Explore how the perimeter changes when the area changes and vice versa

⚡ Calculate the area of parallelograms and triangles

⚡ Calculate and estimate the volume of cubes and cuboids

The regular octagon and regular hexagon have the same perimeter. What is the length of one side of the hexagon?

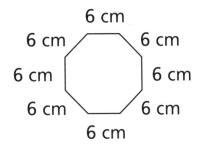

6 cm
6 cm   6 cm
6 cm   6 cm
6 cm   6 cm
6 cm

Here are some maths words we will be using. Which words are new?

area   volume   perimeter

parallelogram   height   enclosed

width   length   square centimetres (cm²)

square metres (m²)   base   estimate

formula   compound shape

cubic centimetres (cm³)   cubic metres (m³)

Describe the pattern. Draw the next shape. Which shape has the largest perimeter? Which has the largest area? How do you know?

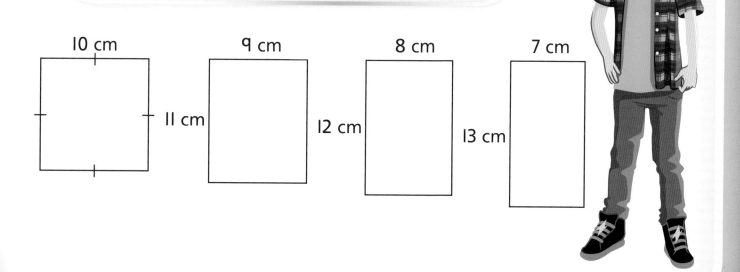

10 cm

11 cm

9 cm

12 cm

8 cm

13 cm

7 cm

Shapes with the same area

Discover

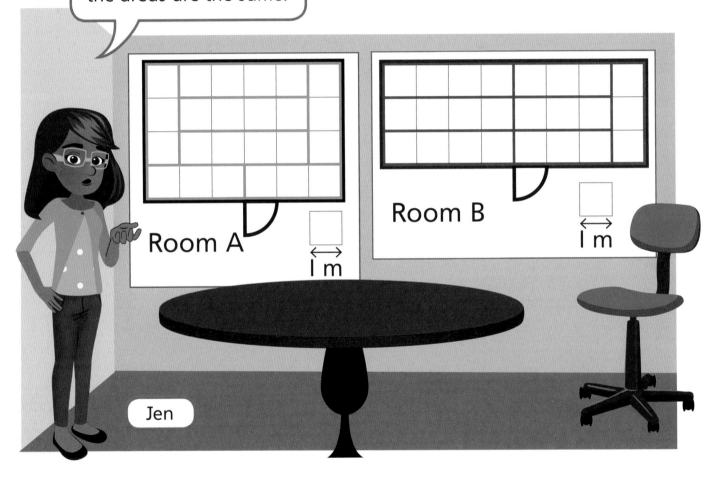

The dimensions of the rooms are different but the areas are the same.

Room A

Room B

I m

I m

Jen

1 **a)** Is Jen correct?

b) Rearrange the parts of one room to make a different shape from the ones shown. What are the dimensions of the new room? What is the area?

Share

a) Both rooms are made of 3 parts that are 4 m long and 4 parts that are 3 m long.

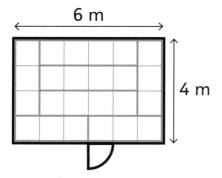

6 m

4 m

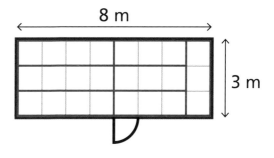

8 m

3 m

Area of Room A
= 4 m × 6 m = 24 m²

Area of Room B
= 3 m × 8 m = 24 m²

Both rooms are rectangles. I used multiplication to work out the area of each room.

The two rooms have different dimensions, but their areas are equal.

Jen is correct.

b) This example is made of the same parts as Room A and Room B. The dimensions of this room are 2 m wide and 12 m long.

The dimensions of the room are its width and length.

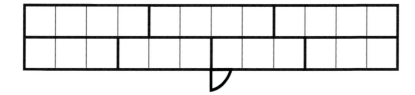

The area is 2 m × 12 m = 24 m². This room has the same area as Room A and Room B.

Think together

1 Find the shapes that have the same area.

Shape A area = ☐ m²

Shape B area = ☐ m²

Shape C area = ☐ m²

Shape D area = ☐ m²

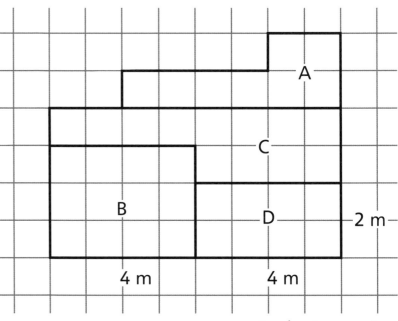

Scale: 1 cm = 1 m

Shapes ☐ and ☐ and shapes ☐ and ☐ have the same area.

2 a)

These two shapes have the same area.

Luis

No, they do not because one of the shapes looks smaller.

Lexi

4 cm

9 cm

1 cm ↔

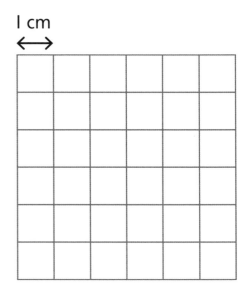

Who do you agree with? Explain why.

b) On squared paper, draw one more rectangle that has the same area.

3 Here are three shapes.

A is a square.

B is a rectangle where the length is 4 times as great as the width.

C is a compound shape.

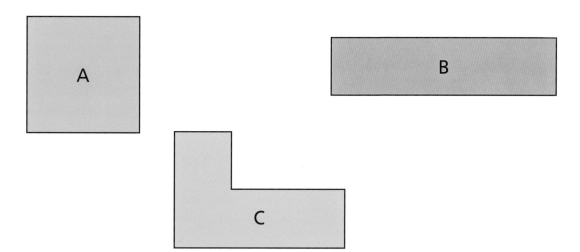

a) The area of each shape is 16 cm². Work out the dimensions of each shape.

b) On paper, accurately draw each shape.

A **compound shape** is a shape that is made of two or more shapes put together, like shape C.

To work out the length of the rectangle I will use a bar model to help me.

→ Practice book 6B p117

Area and perimeter ❶

Discover

Lexi

Max

❶ a) What is the same about the shapes Max and Lexi have drawn?
What is different?

b) Max draws another rectangle on his grid. It has the same area but a
different perimeter. What could the dimensions of the shape be?

Share

a) Lexi's shape is a square.

The length of each side is 6 cm.

The perimeter of the square is 4 × 6 cm = 24 cm.

The area of the square is 6 × 6 cm = 36 cm².

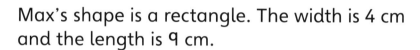

Max's shape is a rectangle. The width is 4 cm and the length is 9 cm.

The perimeter is 2 × 4 cm + 2 × 9 cm

= 8 cm + 18 cm = 26 cm.

The area is 4 cm × 9 cm = 36 cm².

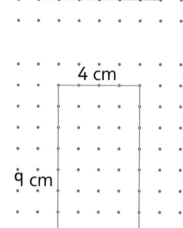

Both shapes have the same area but different perimeters.

b) Max draws a rectangle with an area of 36 cm².

Max's rectangle could be 12 cm long and 3 cm wide.

Area = 12 cm × 3 cm = 36 cm²

Perimeter = 12 cm × 2 + 3 cm × 2 = 30 cm

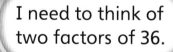

I need to think of two factors of 36.

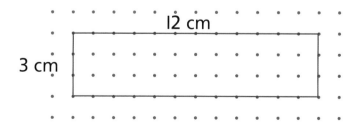

Think together

1 **a)** Work out the area and the perimeter of each shape below. What are the missing numbers in the table?

A

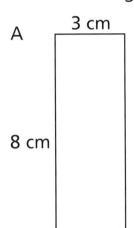

3 cm

8 cm

B

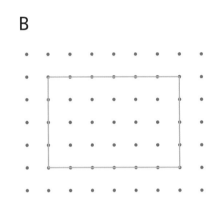

C

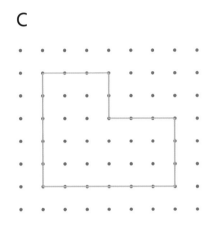

Shape	Area	Perimeter
A	☐ × ☐ = ☐	☐ × 2 + ☐ × 2 = ☐
B		
C		

b) Draw or make another shape with the same area but a different perimeter.

2 Look at these shapes.

What do you notice about the area and perimeter of each of the shapes?

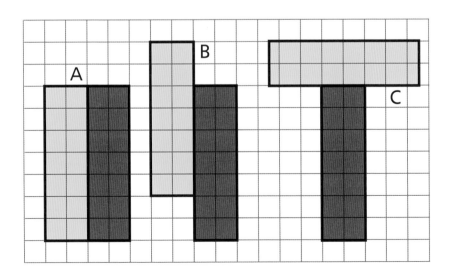

166

3 Is Luis correct?

I think that if the area of a shape becomes smaller, then the perimeter becomes smaller. If the area becomes larger, then the perimeter becomes larger.

Luis

Explain your answer.

I will start by drawing a square. Then I will change the area to see what happens.

I will draw two rectangles and put measurements on. Then I will see if I can find an example that does not work.

167

Area and perimeter 2

Discover

We used ribbon to decorate the edges of the cards. We used 32 cm of ribbon for each card.

Your card is a square and mine is a rectangle. The length of my card is 10 cm.

Lee

Jamilla

1 The ribbon fits around each of the cards exactly.

a) What is the area of Lee's card? What is the area of Jamilla's card?

b) Mrs Dean picks up another card from the table. What could be the area of the card she chooses? What do you notice?

Share

a) Find the length and width of each card, then find the area.

The length of each ribbon is 32 cm. So the perimeter of each card is 32 cm.

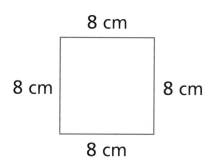

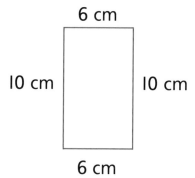

Length of the square

$= 32 \div 4 = 8$ cm

Area $= 8$ cm $\times 8$ cm $= 64$ cm^2

The area of Lee's card is 64 cm^2.

Length + width $= 32 \div 2 = 16$ cm

Width $= 16$ cm $- 10$ cm $= 6$ cm

Area $= 6$ cm $\times 10$ cm $= 60$ cm^2

The area of Jamilla's card is 60 cm^2.

b) Perimeter $= 32$ cm so length + width $= 16$ cm.

Some possible areas of Mrs Dean's card are:

Area of A $= 13 \times 3 = 39$ cm^2

Area of C $= 11 \times 5 = 55$ cm^2

Area of B $= 12 \times 4 = 48$ cm^2

Area of D $= 7 \times 9 = 63$ cm^2

I notice that rectangles can have the same perimeter, but different areas.

Think together

1 Three different gardens each have a perimeter of 20 m.

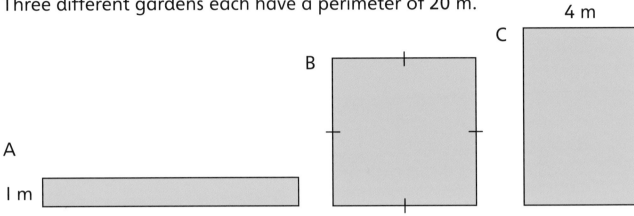

A

1 m

B

C

4 m

a) What are the missing lengths and widths?

b) What is the area of each garden?

c) What is the same about the gardens? What is different?

2 Calculate the areas and perimeters of these pairs of shapes, then compare them.

a)

A

3 cm

5 cm

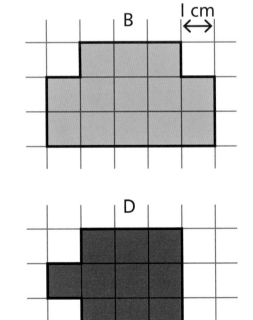

B

1 cm

b)

C

3 cm

4 cm

D

3 A rectangle has a perimeter of 30 m.

a) What dimensions could the rectangle have if the width is a prime number? Write your results in a table.

I think the width has to be shorter than the length.

I think that the width is a prime number less than 15.

b) Find the rectangle with the smallest area and the rectangle with the largest area.

c) A rectangle has a perimeter of 40 cm. Isla says the length of the rectangle has to be less than 20 cm. Is Isla correct?

→ **Practice book 6B p123**

Area of a parallelogram

Discover

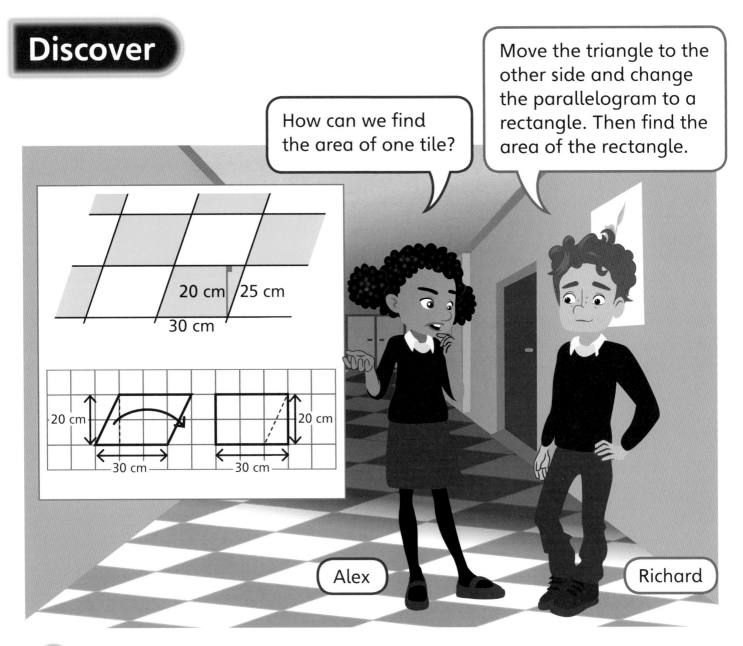

How can we find the area of one tile?

Move the triangle to the other side and change the parallelogram to a rectangle. Then find the area of the rectangle.

Alex

Richard

1 The floor is made from tiles that are parallelograms.

a) Use Richard's method to find the area of one tile.

b) Max finds the area of one tile by multiplying 25 cm by 30 cm.

Emma finds the area of the same tile by multiplying 30 cm by 20 cm.

Who will find the correct area?

Share

A parallelogram is a quadrilateral. Each pair of opposite sides is parallel.

a) Richard's method changes the parallelogram to a rectangle.

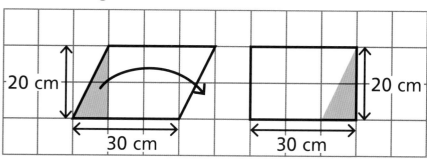

The area of the parallelogram is equal to the area of the rectangle.

The area of the parallelogram = 30 cm × 20 cm = 600 cm^2.

So the area of one tile is 600 cm^2.

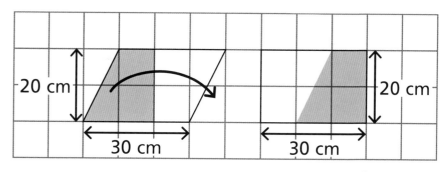

I can do it differently, but get the same result.

b) Max's method is 25 cm × 30 cm = 750 cm^2.

Max is incorrect.

Emma's method is 30 cm × 20 cm = 600 cm^2.

Emma will find the correct area.

Area of a parallelogram = base × height

In a parallelogram, the perpendicular distance from the base to the top is called the perpendicular height.

Think together

1 For i) find the area of each parallelogram by changing it to a rectangle.

For ii) use this formula to calculate the area of each parallelogram:

Area = base × height

a) Area of parallelogram A

   i) Area of the parallelogram = area of the rectangle with width ☐ cm and length ☐ cm.

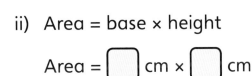

Area = ☐ cm²

   ii) Area = base × height

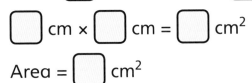

Area = ☐ cm²

b) Area of parallelogram B

   i) ☐ × ☐ = ☐ cm²

   ii) Area = ☐ × ☐

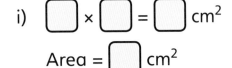

Area = ☐ cm²

c) Area of parallelogram C

   i) ☐ × ☐ = ☐

Area = ☐ cm²

   ii) Area = ☐ × ☐

Area = ☐ cm²

1cm

174

2 Find the area of each of these parallelograms.

a)

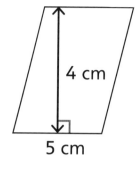

4 cm

5 cm

c)

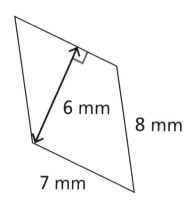

6 mm

8 mm

7 mm

b)

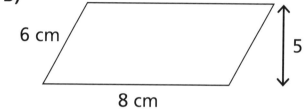

6 cm

5 cm

8 cm

d)

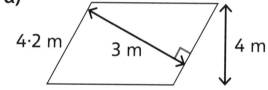

4·2 m

3 m

4 m

3 Find the missing measurements.

Area = 24 cm²

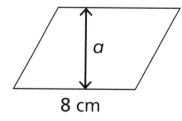

a

8 cm

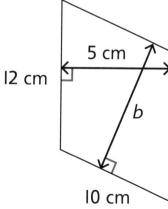

5 cm

12 cm

b

10 cm

CHALLENGE

175

Area of a triangle ❶

Discover

❶ a) Work out the area of Andy's triangle.

b) What is the area of Jamie's triangle?

Share

a) **Method I**

> I estimated the area by counting the squares.

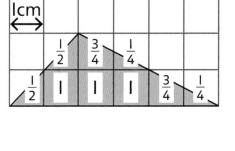

$1 \times 3 + \frac{3}{4} \times 2 + \frac{1}{2} \times 2 + \frac{1}{4} \times 2$

$= 3 + 1\frac{1}{2} + 1 + \frac{1}{2}$

$= 6$ squares or 6 cm^2

Method 2

Change the triangle to a rectangle by moving some part squares.

The area of the triangle is the same as the area of the rectangle: $6 \times 1 = 6$ cm^2.

The area of Andy's triangle is 6 cm^2.

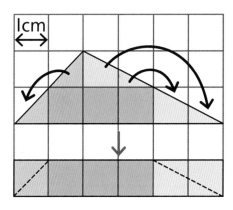

b) Change Jamie's triangle to a rectangle by splitting it exactly in half and moving one triangle to the other side.

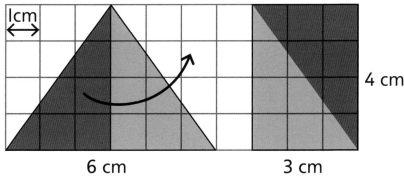

6 cm   3 cm   4 cm

Area of the rectangle = 4 cm × 3 cm = 12 cm^2.

The area of Jamie's triangle is 12 cm^2.

Think together

1 Estimate the areas of the triangles below.

a)

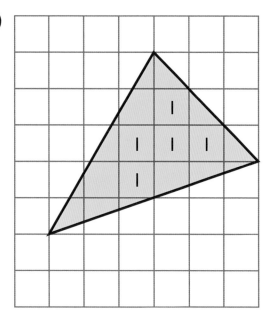

b)

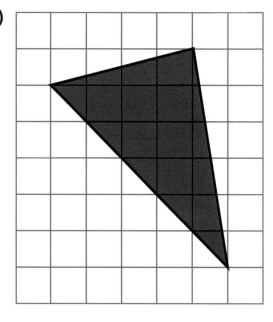

2 Find the areas of the following triangles by making rectangles.

a)

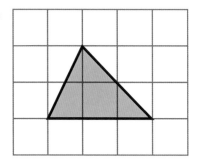

b)

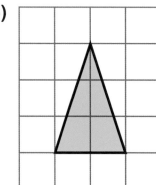

3 Olivia estimates the area of the triangle by counting the squares. She makes a table to record her results.

Complete Olivia's table.

Whole squares	
Almost-whole squares	
Half squares	
Quarter squares	
Less than a quarter squares	

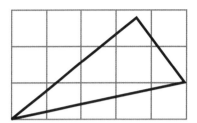

Andy works out the area by changing it to two rectangles.

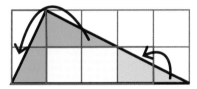

Use both methods to work out the area of the triangle.

I wonder which method is more accurate.

179

Area of a triangle ②

Discover

Reena

Ebo

1 a) Ebo has cut a rectangle in half.

What do you notice about the area of the rectangle and the area of the triangles?

b) What is the area of Reena's triangle?

Share

a)

I can cut a rectangle into two right-angled triangles.

If I turn one of the triangles around, they look exactly the same.

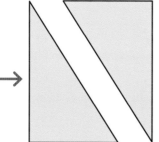

The area of the rectangle is twice the area of each triangle.

b) The area of Reena's triangle is half the area of the rectangle.

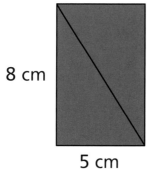 → 8 cm

8 cm

5 cm   5 cm   5 cm

Area of the triangle = area of the rectangle ÷ 2

Area of the rectangle = 8 cm × 5 cm = 40 cm²

Area of the triangle = 40 cm² ÷ 2 = 20 cm²

The area of Reena's triangle is 20 cm².

Think together

1   Calculate the area of each rectangle and each shaded triangle.

   a) Area of rectangle = ⬚ cm × ⬚ cm = ⬚ cm²

      Area of triangle = ⬚ cm² ÷ 2 = ⬚ cm²

   b) Area of rectangle = ⬚ × ⬚ = ⬚ cm²

      Area of triangle = ⬚ ÷ 2 = ⬚ cm²

   c) Area of rectangle = ⬚ × ⬚ = ⬚ cm²

      Area of triangle = ⬚ ÷ 2 = ⬚ cm²

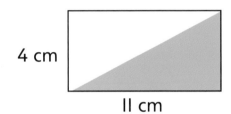

3 cm, 6 cm

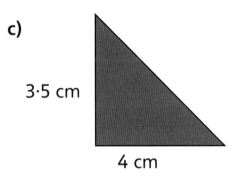

2 cm, 2 cm, 4 cm, 11 cm

2   Find the area of each triangle below.

   a)

   4 cm

   4 cm

   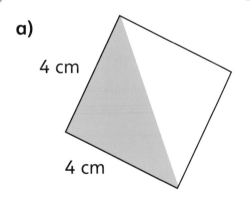

   b)

   2·5 cm

   4 cm

   c)

   3·5 cm

   4 cm

   d)

   8 cm

   5·5 cm

   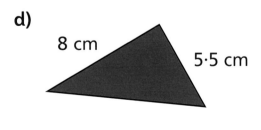

3 AD is the base of the triangle ADC. CD is the height of the right-angled triangle.

a) Here is Olivia's method for calculating the area of the right-angled triangle. Is she correct?

Area of the triangle = Area of the rectangle ÷ 2, so the area of the triangle = base × height ÷ 2.

Olivia

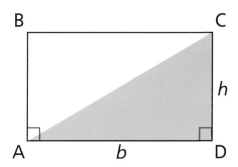

We use *b* for base and *h* for height.

b) What could the base and height of a triangle be if the area is 10 cm²?

I think I need to find two numbers that multiply together to make 20.

183

→ **Practice book 6B p132**

Area of a triangle ③

Discover

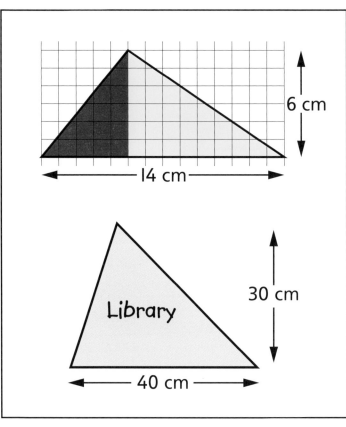

Mo

Lexi

1 **a)** What is the area of the triangle on Lexi's sign?

b) What is the area of Mo's sign?

Share

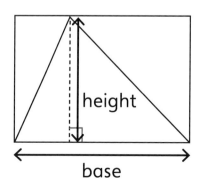

Area of rectangle =
2 × area of triangle

Area of triangle = area
of rectangle ÷ 2

a) Area of rectangle = 14 × 6 = 84 cm²

Area of triangle = 84 ÷ 2

The area of the triangle on Lexi's sign is 42 cm².

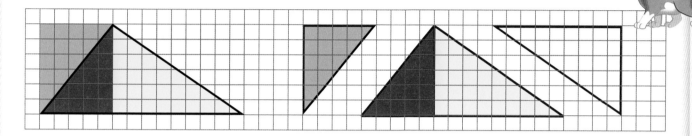

I found the areas of the red and
yellow triangles and added them,
but I think there is a quicker way.

This method is quicker!

b) The area of Mo's sign is half the area of
the rectangle.

Area of rectangle = 40 × 30 = 1,200 cm²

Area of triangle = 1,200 ÷ 2 = 600 cm²

The area of Mo's sign is 600 cm².

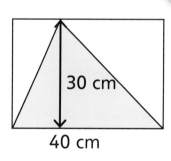

30 cm

40 cm

For any triangle we can say that:
Area of triangle = base × height ÷ 2.

The height is always perpendicular
to the base.

height

base

185

Think together

1 Calculate the area of each triangle.

a)

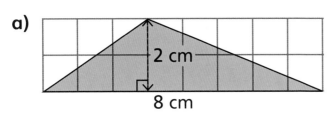

Area = ☐ × ☐ ÷ 2

Area = ☐ cm²

b)

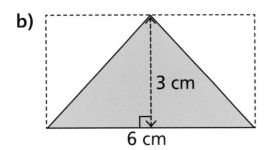

Area = ☐ × ☐ ÷ 2

Area = ☐ cm²

c)

4 cm

11 cm

Area = ☐ × ☐ ÷ 2

Area = ☐ cm²

2

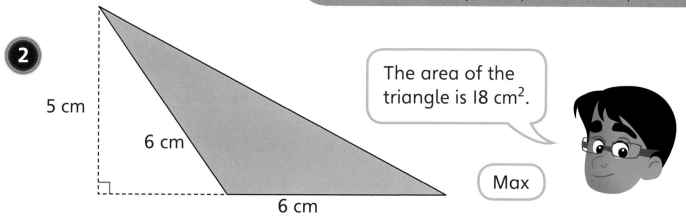

5 cm

6 cm

6 cm

The area of the triangle is 18 cm².

Max

a) What mistake has Max made? What should he have done instead?

b) Measure the base and the height of the triangle below and find its area.

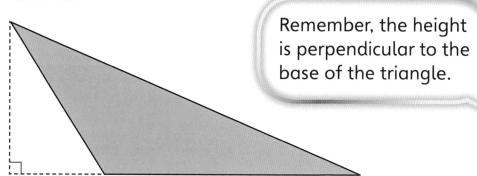

Remember, the height is perpendicular to the base of the triangle.

3 a) Which of these triangles has the larger area?
Explain your reasoning.

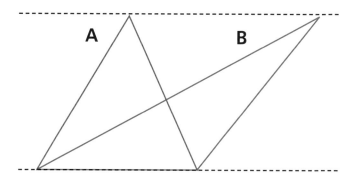

A

B

CHALLENGE

I think there is a way to explain without actually calculating the areas.

b) On your own paper, draw another triangle that has a larger area than triangle A.

187

→ Practice book 6B p135

Problem solving – area

Discover

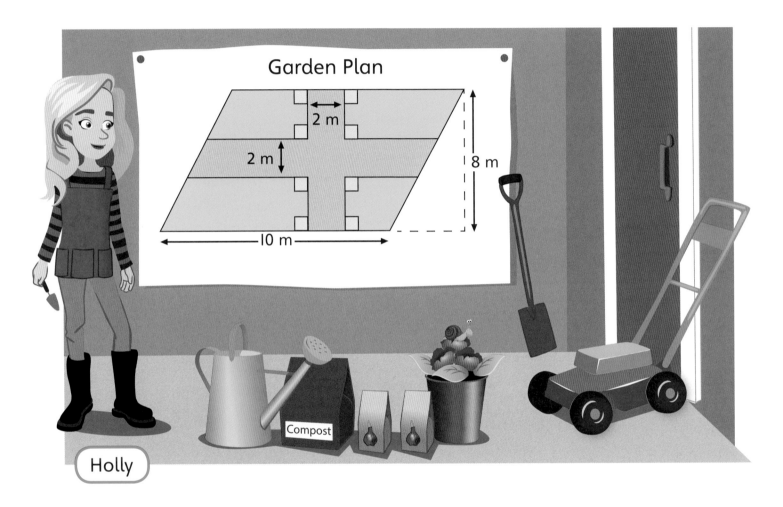

Garden Plan

2 m

2 m

8 m

10 m

Holly

1 The garden has two paths that are perpendicular to each other. Each path is 2 m wide.

a) What is the total area of the paths?

b) What is the total area of the grass?

Share

a) This path is the shape of a rectangle.

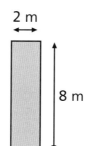

2 m

8 m

This path is the shape of a parallelogram.

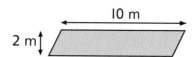

10 m

2 m

Area of the rectangle
= 8 × 2 = 16 m²

Area of the parallelogram
= 10 × 2 = 20 m²

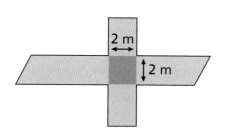

2 m

2 m

> The paths overlap! I wonder how I can find the total area without including the area of the overlapping square twice?

To find the total area of the paths, add up the area of each path and subtract the area of the square.

Area of the square = 2 × 2 = 4 m²
16 + 20 = 36 m²
36 − 4 = 32 m²
The total area of the paths is 32 m².

b) To find the area of the grass, find the area of the parallelogram then subtract the area of the paths.
Area of the parallelogram = 10 × 8 = 80 m²
Area of the paths = 32 m²

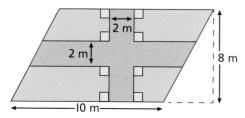

2 m

2 m

8 m

10 m

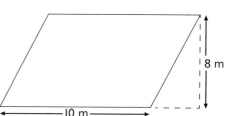

8 m

10 m

Area of the grass = 80 − 32 = 48 m²
The total area of the grass is 48 m².

Think together

1 The school garden is a rectangular shape. Use both Isla's and Mo's methods to find the total area of the garden that is made up of flowers.

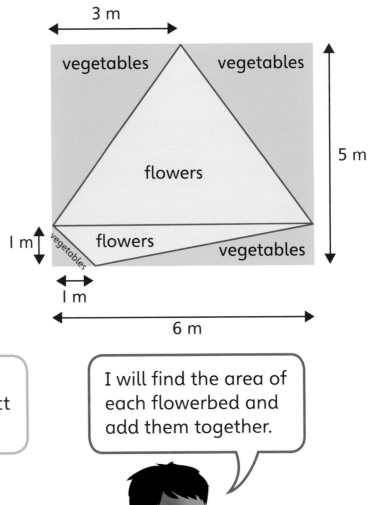

I will find the area of the whole garden, then subtract the area with vegetables.

Isla

I will find the area of each flowerbed and add them together.

Mo

2 Sofia is calculating the number of triangular tiles needed to cover her bathroom floor.

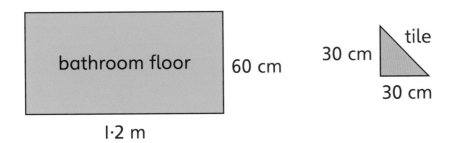

a) What is the area of the floor?

b) What is the area of one tile?

c) How many triangular tiles are needed to cover the whole floor?

3 Toshi's garden is a rectangle of grass surrounded by a path 1 m wide.

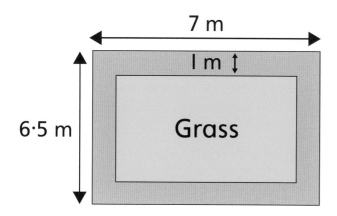

7 m

1 m ↕

6·5 m

Grass

a) Calculate the area of the path.

b) Toshi wants to test out his new tent by putting it up on the grass. The tent will take up 4·5 m².

Is there enough room on the grass for the tent?

I think I can find the solution in two ways.

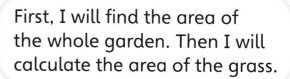

First, I will find the area of the whole garden. Then I will calculate the area of the grass.

191

Problem solving – perimeter

Discover

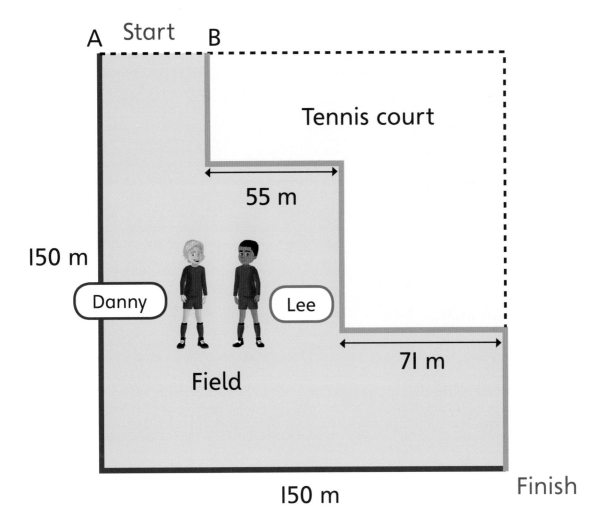

1 Danny and Lee were running around the park. Danny took route A and Lee took route B.

a) Who ran the greater distance? How much further did they run?

b) The running club want to hang bunting either around the perimeter of the park or around the perimeter of the field.

Which option uses less bunting?

Share

a) Danny ran along 2 straight paths:
150 m + 150 m = 300 m

Lee ran along 2 paths of 55 m and 71 m:
55 + 71 = 126 m

Lee also ran three shorter paths.

126 + 150 = 276 m

Lee ran 276 m. Danny ran 300 m.

Danny ran the greater distance.

300 – 276 = 24

Danny ran 24 metres further than Lee.

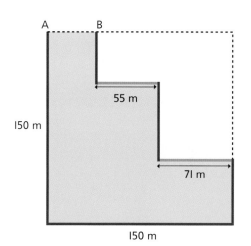

The three shorter paths must add up to 150 m because they are the same length as one side of the square.

b) The perimeter of the park = 150 m × 4 = 600 m.

To find the perimeter of the field, add the distances run by Danny and Lee and the distance between the start points A and B.

The distance between A and B is
150 – 55 – 71 = 24 m.

The perimeter of the field is
300 + 276 + 24 = 600 m.

The perimeter of the field is the same as the perimeter of the park.

The amount of bunting will be the same wherever the running club choose to put it.

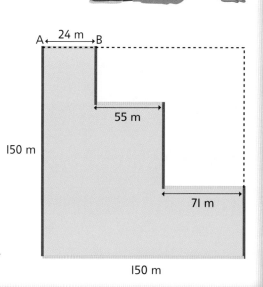

Think together

1 Max has a square piece of material. The perimeter of the material is 40 cm.

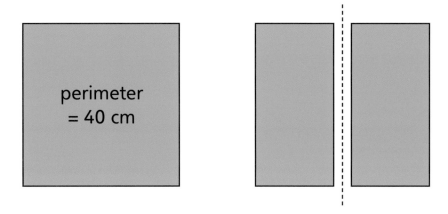

perimeter
= 40 cm

He cuts the material into two equal rectangular pieces.

What is the perimeter of each rectangle?

2 The regular hexagon and regular octagon below have the same perimeter.

If the length of one side of the octagon is 9 cm, what is the length of one side of the hexagon?

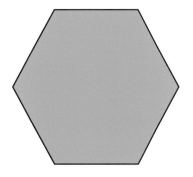

9 cm

3 The area of the small square is 25 cm². The area of the large square is 81 cm². Find the perimeter of the unshaded shape.

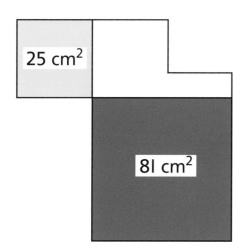

25 cm²

81 cm²

CHALLENGE

4 Amal has a rope that is 48 m long.

He puts the rope around a rectangular patch of land where seedlings have been planted.

The width and the length of the rectangle are whole numbers.

What is the greatest possible area of the patch of land?

I think the length and width of the rectangle must add up to 24.

I wonder what the length and width could be.

195

Volume of a cuboid ❶

Discover

My solid has a greater volume than yours. It takes up more space.

We are both using cubes that each have a volume of 1 **cm³**.

Jamilla

Max

❶ a) Is Jamilla correct?

b) The children put the cubes from both of their solids together to make a cuboid.

Using cubes, make three different cuboids that have the same volume.

Share

> The amount of space a solid figure takes up is called the **volume** of the solid figure. We can measure volume in cm^3.

a) Work out the volume of each 3D solid by counting the cubes.

Each cube has a volume of I **cubic centimetre** (cm^3).

> I counted the cubes in the solid to work out the volume.

Jamilla's 3D solid has a volume of 8 cm^3.

Max has made a cube. All the sides are the same length.

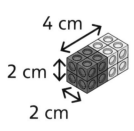

> I thought of the cube as 2 layers of 4 cubes.

$2 \times 2 \times 2 = 4 \times 2 = 8$ cm^3

Jamilla and Max's solids each have a volume of 8 cm^3.
Jamilla is not correct.

b) There are 16 cm^3 in total. Jamilla and Max could make the following cuboids:

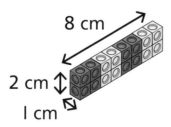
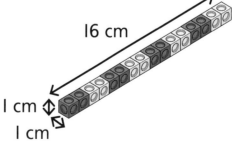

$2 \times 2 \times 4 = 16$ cm^3 $1 \times 2 \times 8 = 16$ cm^3 $1 \times 1 \times 16 = 16$ cm^3

Think together

1 What is the volume of each of the following solids?

a)

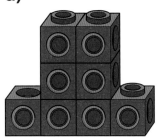

[ ] cm³

b)

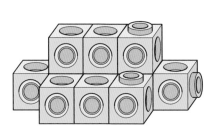

[ ] cm³

c)

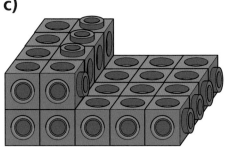

[ ] cm³

2 Work out the volumes of the cuboids.

a)

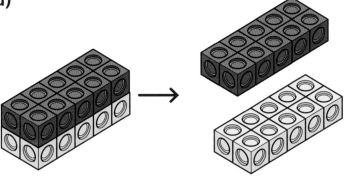

Volume = 2 × 5 × 2

= [ ] × 2

= [ ] cm³

The volume of the cuboid

is [ ] cm³.

b)

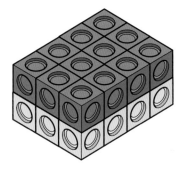

Volume = [ ] × [ ] × [ ]

= [ ] × [ ]

= [ ]

The volume of the cuboid is [ ] cm³.

3 **a)** Max and Jamilla are working out the volume of this cuboid using multiplication.

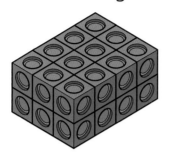

They split the shape into parts to help them.

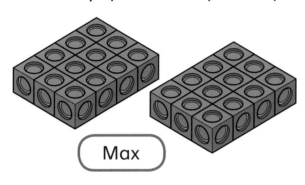

 Max

 Jamilla

What calculation did each child use to find the volume?

b) Which of these 3D shapes has the greatest volume?

A   B   C

I will use multiplication to work out each volume.

I wonder if I can arrange all the small cubes to make a larger cube. Perhaps I can tell by finding the total number of cubes.

199

→ Practice book 6B p144

Volume of a cuboid ❷

Discover

1 **a)** How many I cm³ cubes can fit into each box?

What is the volume of each box?

b) How can you work out the volume of each box by multiplying?

Share

a) Find the number of I cm³ cubes that can fit into each box. This is the volume.

4 cm
4 cm   4 cm

3 cm
4 cm   5 cm

4 cm
4 cm   4 cm

3 cm
4 cm   5 cm

> I put cubes into each box until it was full. Then I counted the number of cubes. It took me a long time.

4 × 4 = 16

16 × 4 = 64 cubes

The volume of Aki's box is 64 cm³.

5 × 4 = 20 cubes on the bottom layer

20 × 3 layers = 60 cm³

The volume of Kate's box is 60 cm³.

> I worked out how many I cm³ cubes fit in the bottom layer and then multiplied by the number of layers. This is the same as multiplying length × width × height.

> To find the volume of a cuboid, I can work out length (l) × width (w) × height (h).

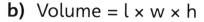

b)
Volume = l × w × h
= 4 × 4 × 4
= 16 × 4
= 64 cm³

Volume = l × w × h
= 5 × 4 × 3
= 20 × 3
= 60 cm³

> We can write the formula for volume like this.

Think together

1 Work out the volume of each of the 3D solids.

a)

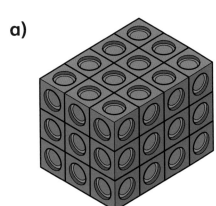

Volume = 4 × 3 × 3

= ☐ × 3

= ☐ cm³

b)

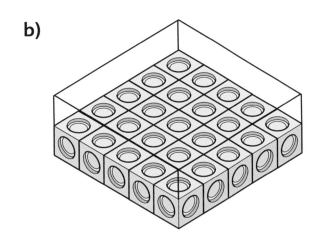

Volume = 5 × 5 × 2

= ☐ × ☐

= ☐ cm³

c)

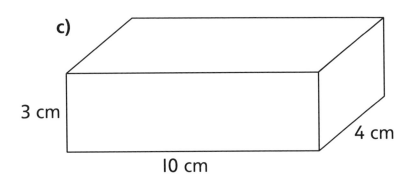

3 cm

4 cm

10 cm

Volume = ☐ × ☐ × ☐

= ☐ × ☐

= ☐ cm³

2 Olivia thinks the volume of each of these solids is the same.
Do you agree?

Write down how you would explain this to a partner.

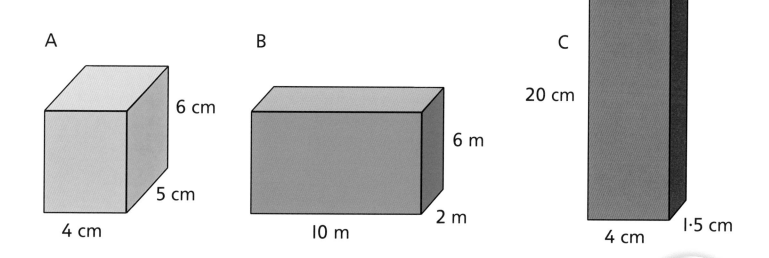

A

6 cm

5 cm

4 cm

B

6 m

2 m

10 m

C

20 cm

4 cm

1·5 cm

3 **a)** This cuboid has a volume of 40 cm³.
Find the missing height.

CHALLENGE

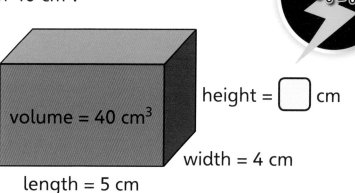

volume = 40 cm³

length = 5 cm

height = ☐ cm

width = 4 cm

b) A cube has a volume
of 1,000 cm³. Find the
missing length.

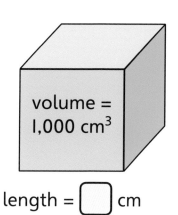

volume =
1,000 cm³

length = ☐ cm

203

→ **Practice book 6B p147**

End of unit check

1 The area of the parallelogram and the area of the triangle are equal. What is the length of the base of the triangle?

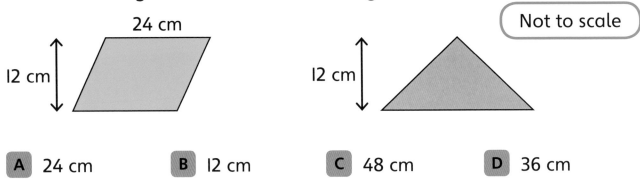

24 cm

12 cm

Not to scale

12 cm

A 24 cm   B 12 cm   C 48 cm   D 36 cm

2 Which of these four statements is true?

A To find the area of a rectangle, multiply the length by the width, then double the answer.

B To find the area of a triangle, measure the length of a side, then multiply it by 3.

C To find the area of a parallelogram, multiply the width by the length.

D To find the volume of a cuboid, multiply length by width by height.

3 The volumes of cube A and cuboid B are equal. What is the missing length?

A 18 cm   C 9 cm

B 3 cm   D 8 cm

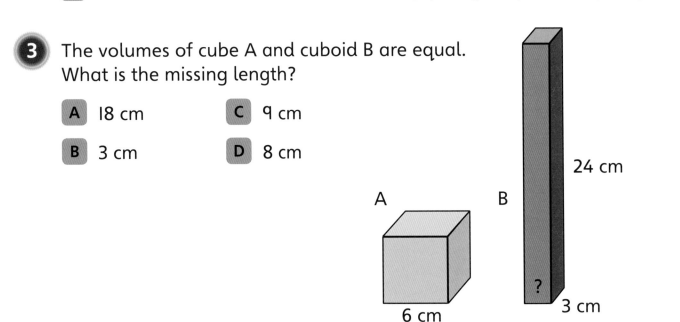

A

B

24 cm

?

6 cm

3 cm

4 Here are some shapes on a centimetre grid. Which of these four statements is true?

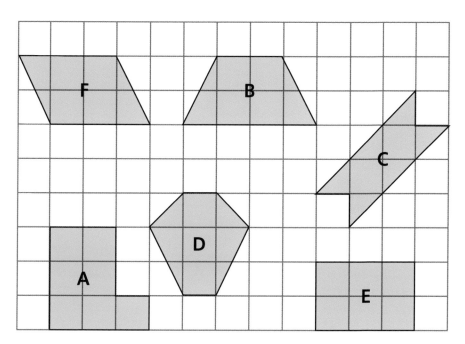

A Shapes A and E have the same areas but different perimeters.

B Shapes B and F have the same perimeters and areas.

C Shapes C and D have the same perimeters.

D Shapes A and E have equal perimeters but different areas.

5 This triangle shows a lawn. Turf costs £2 per m².

a) What is the area of the lawn?

b) How much would it cost to turf the lawn?

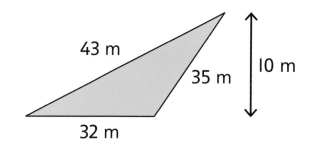

43 m 35 m 10 m 32 m

205

→ Practice book 6B p150

Unit 12
Ratio and proportion

In this unit we will …

⚡ Calculate ratios

⚡ Use ratios to work out amounts

⚡ Enlarge shapes by a scale factor

⚡ Identify similar shapes

⚡ Solve problems involving ratio

We will use bar models to represent ratio problems. For every 1 slice of carrot cake there are 4 slices of lemon cake. If there are 20 slices in total, how many slices are carrot?

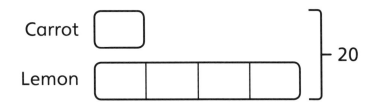

Carrot

Lemon

20

We will need some maths words. We will also often be using the phrase, 'For every ... there are ...'. What do you think it might mean?

ratio proportion part

whole scale scale factor

similar notation

We will need to know our multiplication and division facts. Write three multiplication or division facts that match this one.

$$8 \times 9 = \boxed{}$$

Ratio ❶

Discover

We need to sort ourselves into equal groups for the walk.

Mrs Dean

1 a) Sort the children and adults into equal groups.

Use the groups you have made to complete the sentence.

For every ⬚ adult there are ⬚ children.

b) Write a sentence to compare how many bananas and how many apples there are.

Share

a) There are 2 adults and 6 children.

> I sorted the people into equal groups. Each group has 1 adult and 3 children.

For every 1 adult there are 3 children.

> This statement is an example of a **ratio**. A ratio compares two or more parts of the whole.

b) In each group there are 3 bananas and 2 apples.

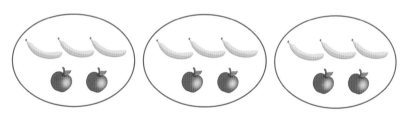

So, for every 3 bananas there are 2 apples.

Or for every 2 apples there are 3 bananas.

> I tried to sort the fruit so there was 1 apple in each group, but I did not have enough bananas.

Think together

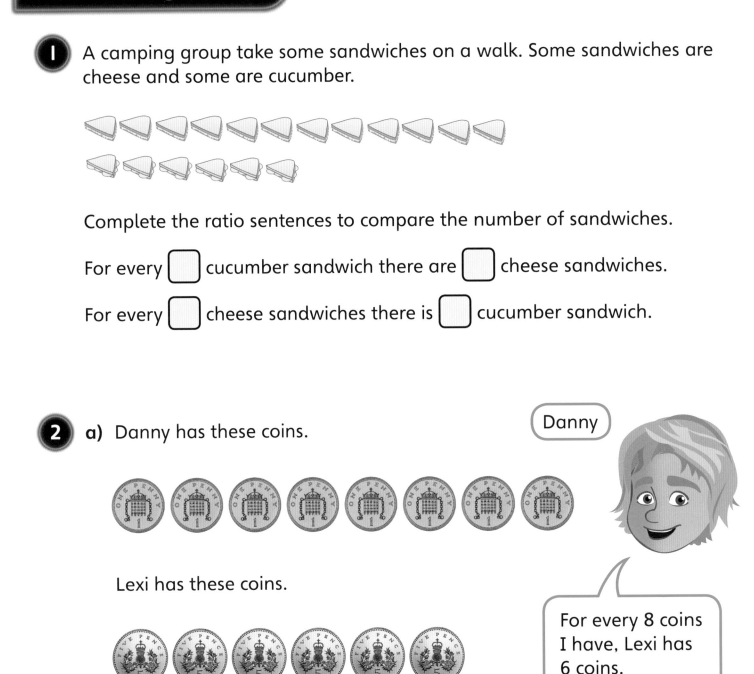

1   A camping group take some sandwiches on a walk. Some sandwiches are cheese and some are cucumber.

Complete the ratio sentences to compare the number of sandwiches.

For every ☐ cucumber sandwich there are ☐ cheese sandwiches.

For every ☐ cheese sandwiches there is ☐ cucumber sandwich.

2   a) Danny has these coins.

Danny

Lexi has these coins.

For every 8 coins I have, Lexi has 6 coins.

Is Danny correct?

b) Write another ratio sentence to compare the coins that Lexi and Danny have.

3 **a)** Josh is making towers of cubes.

What is the same about the two towers?

What is different?

b) Josh makes another tower.

For every 3 red cubes there are 2 yellow cubes.

What fraction of the tower is red?

What fraction of the tower is yellow?

I worked out what fraction of each tower was red. I noticed something about the answers.

I wrote some ratio sentences. I wonder if there is a link between these and your fraction answers.

→ **Practice book 6B p153**

Ratio ❷

Discover

On my next t-shirt, for every I star I am going to print 3 suns.

Emma

T-shirt printing

1 a) Complete the ratio sentence.

On Emma's first completed t-shirt, for every ⬜ stars there are ⬜ suns.

b) Emma makes her next t-shirt.

What could the t-shirt look like?

Share

a) There are 6 stars and 9 suns.

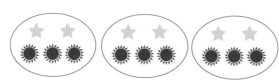

6 stars : 9 suns
÷ 3   ÷ 3
2 stars : 3 suns

> I have written my answer in its simplest form. I noticed that 6 and 9 both divide by 3.

On Emma's first completed t-shirt, for every 2 stars there are 3 suns.

> We can say and write this differently. The ratio of stars to suns is 2 to 3. We write this as 2 : 3.

b) On Emma's next t-shirt, for every 1 star there are 3 suns.

Emma's t-shirt could look like this:

> For every 1 star I drew, I had to draw 3 suns. The t-shirt has three times as many suns as stars on it.

Stars	1	2	3
Suns	3	6	9
	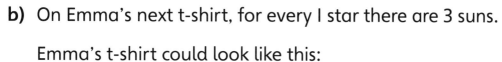		

We can say the ratio of stars to suns is 1 : 3.

213

Think together

1 What is the ratio of squares to circles on each of these t-shirts?

a)

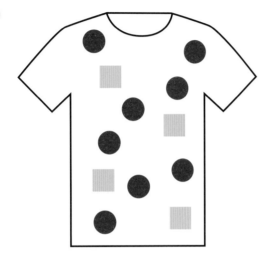

For every ⬜ square there are ⬜ circles.

Or, the ratio of squares to circles is ⬜ : ⬜ .

I remember I read the ':' as 'to'.

b)

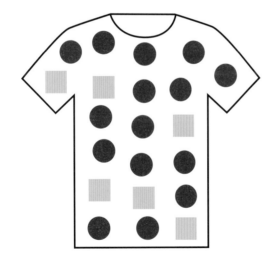

For every ⬜ squares there are ⬜ circles.

Or, the ratio of squares to circles is ⬜ : ⬜ .

Try to write your answers in their simplest form.

2 The ratio of trucks to cars in a car park is 2 : 1.

That means there are more trucks than cars.

Jamilla

This means there are three times as many trucks as cars.

Lexi

There could be 24 trucks and 12 cars.

Zac

Who is correct? Explain your answer.

3 What is the same about the rectangles? What is different about them?

CHALLENGE

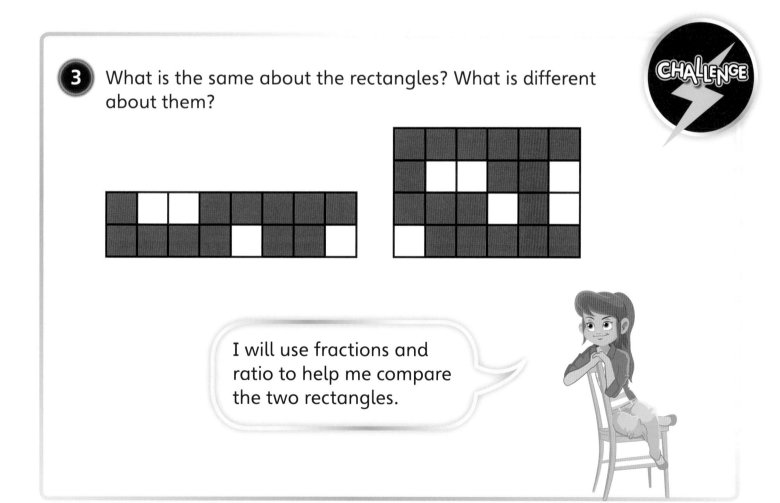

I will use fractions and ratio to help me compare the two rectangles.

→ Practice book 6B p156

Ratio ③

Discover

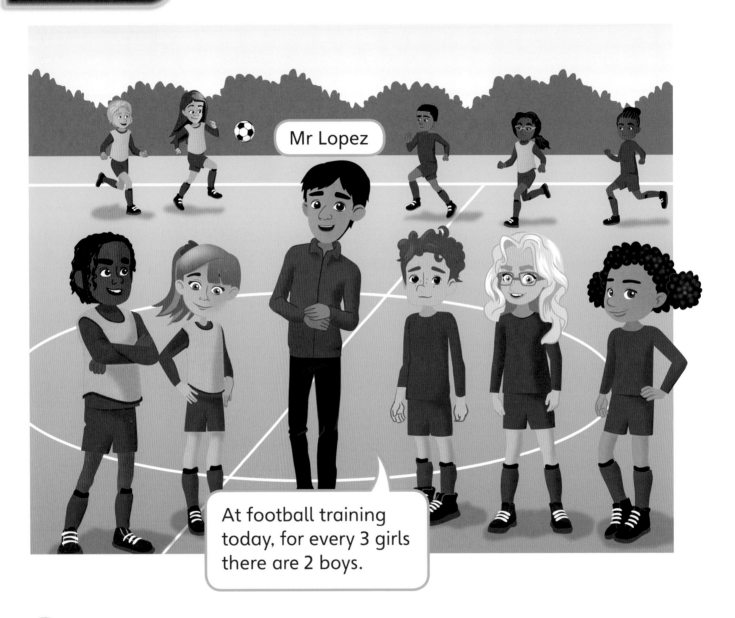

At football training today, for every 3 girls there are 2 boys.

1 a) If there are 12 girls at football training today, how many boys are there?

b) What fraction of the children at football training are girls?

What fraction are boys?

216

Share

a) There are 3 girls for every 2 boys.

Method I

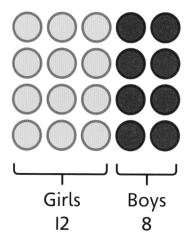

Girls	Boys
12	8

I used counters to represent the children. I placed the counters in rows until I had 12 girls.

I tried two ways! I made a table and added on 3 girls and 2 boys each time. I stopped when I got to 12 girls and then saw there were 8 boys.

Method 2

Add 3 girls each time.

Stop when you get to 12 girls.

Girls	Boys
3	2
6	4
9	6
12	8

Add 2 boys each time.

Method 3

$3 \times 4 = 12$

To calculate the number of boys, work out
$2 \times 4 = 8$

There are 8 boys at football training today.

I also tried it this way: I worked out what I needed to multiply 3 (number of girls) by to make 12. I then multiplied this by 2 to tell me how many boys there are.

b) There are 12 girls and 8 boys. So, there are 20 children in total.

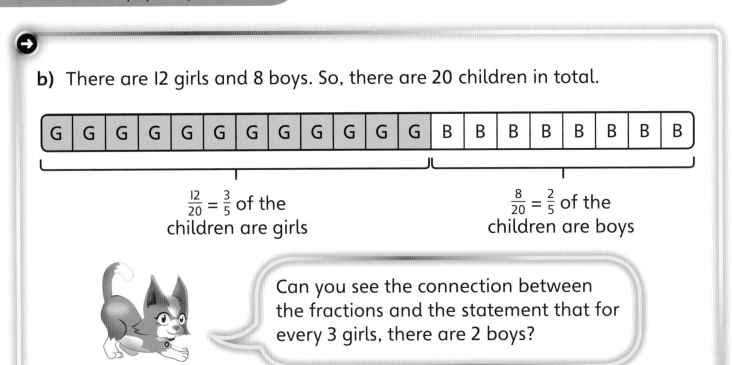

$\frac{12}{20} = \frac{3}{5}$ of the
children are girls

$\frac{8}{20} = \frac{2}{5}$ of the
children are boys

> Can you see the connection between the fractions and the statement that for every 3 girls, there are 2 boys?

Think together

1 For every 2 children wearing a bib, 1 child is not wearing a bib.

If 10 children are wearing a bib, how many children are not wearing one?

Explain your method to a partner.

2 For every I triangle there are 3 squares.

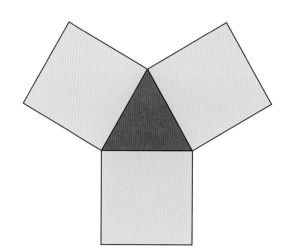

a) If there are 5 triangles, how many squares are there?

b) If there are 18 squares, how many triangles are there?

3 At the park, some children are making patterns.

They make a pattern with leaves and conkers.

They want to continue the pattern.

They have 20 conkers. How many leaves do they need?

I do not think they can use all 20 conkers. I wonder how many of the conkers they can use.

219

Ratio ④

Discover

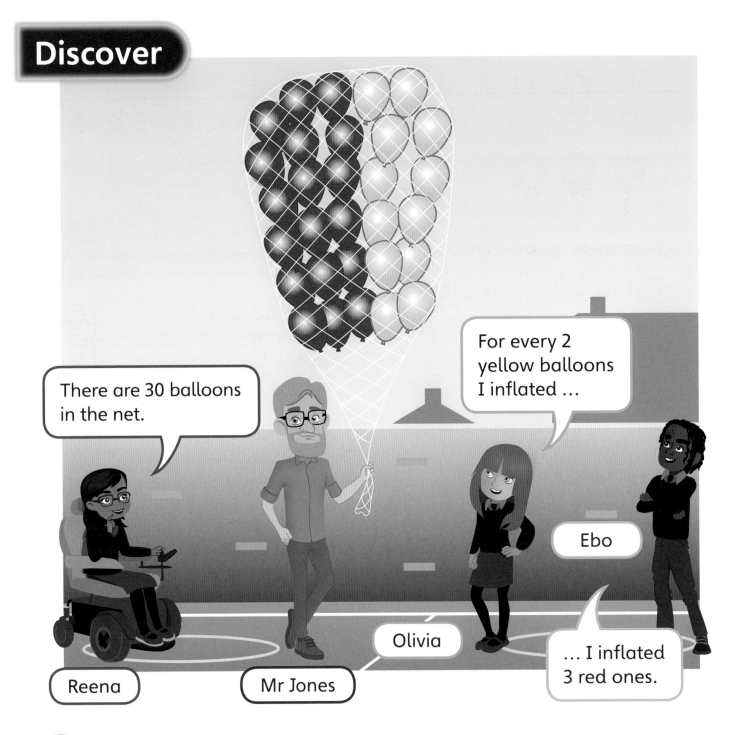

1 a) What is the ratio of yellow to red balloons in the net?

b) How many of the balloons are yellow?

How many are red?

Share

a) For every 2 yellow balloons Olivia inflated, Ebo inflated 3 red ones.

Remember that ratio shows a comparison between two things.

The ratio of yellow to red balloons in the net is 2 : 3.

b) There are 30 balloons in total.

Method 1

I drew balloons in the correct ratio until I reached 30.

There are 5 balloons in each group.

30 ÷ 5 = 6 groups

I worked the answer out. I thought of the 5 balloons as a group and then worked out how many groups of balloons there are in 30.

There are 2 yellow balloons in each group.

6 × 2 = 12

There are 3 red balloons in each group.

6 × 3 = 18

There are 12 yellow balloons and 18 red balloons in the net.

Method 2

Make a table using the ratio of yellow to red balloons.

For each new row, add 2 yellow and 3 red balloons and add up the total number of balloons. Stop when the total reaches 30.

There are 12 yellow balloons and 18 red balloons in the net.

Yellow	Red	Total
2	3	5
4	6	10
6	9	15
8	12	20
10	15	25
12	18	30

Think together

1   There are 21 children in a karate club.

For every 1 girl in the club, there are 2 boys.

How many girls are there in the karate club?

How many boys are there?

2 Classes 6A and 6B raise some money for charity.

For every £2 that Class 6A raises, Class 6B raises £5.

In total they raise £154.

How much money does each class raise?

> For every 2 pound coins I draw for 6A, I am going to draw 5 pound coins for class 6B. I will keep doing this until I draw 154 coins.

> This will take a long time. I think there is a more efficient way.

CHALLENGE

3 Max used a bar model to help him solve the question in **Discover**.

a) Explain how Max used a bar model to help him.

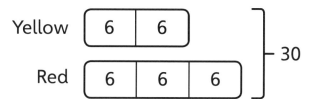

Yellow | 6 | 6
Red | 6 | 6 | 6
} 30

> This shows that for every 2 yellow balloons there are 3 red balloons. It shows that there are 30 balloons altogether.

> There are 5 parts in total, so I can work out 1 part by dividing 30 by 5.

Max

b) Draw bar models to help you solve questions 1 and 2 in **Think together** again.

Did you find them useful?

223

→ **Practice book 6B p162**

Scale drawings

Discover

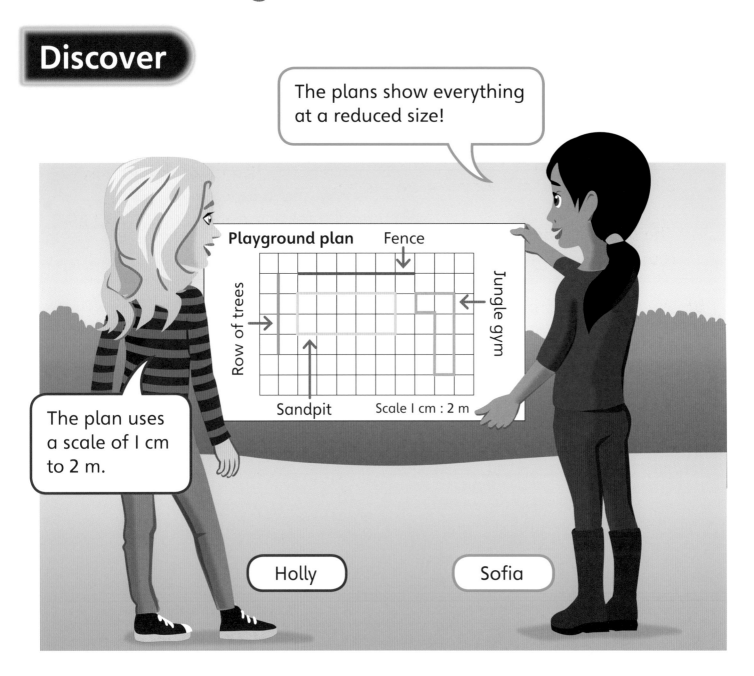

The plans show everything at a reduced size!

The plan uses a scale of 1 cm to 2 m.

Holly

Sofia

Playground plan Fence

Row of trees

Jungle gym

Sandpit Scale 1 cm : 2 m

1 A new playground is being built.

a) What length would the row of trees be in real life?

b) What is the perimeter of the sandpit in real life?

Share

> Remember, the ratio notation tells you that I cm on the plan represents 2 m in real life.

a) On the plan, the row of trees is 4 cm long.

The plan uses the **scale** I cm : 2 m.

```
0 m    2 m    4 m    6 m    8 m    10 m
├──────┼──────┼──────┼──────┼──────┤
0 cm   I cm   2 cm   3 cm   4 cm   5 cm
```

The row of trees would be 8 m long in real life.

> I can use a map scale like this to show how many metres each centimetre is in real life.

b) We need to work out length and width.

We know that every I cm on the plan is 2 m in real life.

I cm
2 m

The width is 2 cm, which is 2 × 2 = 4 m in real life.

2 cm

I cm	I cm
2 m	2 m

2 × 2 = 4 m

The length is 5 cm, which is 5 × 2 = 10 m in real life.

To find the perimeter of the shape we add up all four sides.

5 cm

I cm	I cm	I cm	I cm	I cm
2 m	2 m	2 m	2 m	2 m

5 × 2 = 10 m

Perimeter = 4 + 10 + 4 + 10 = 28 m

The perimeter of the sandpit will be 28 m in real life.

Think together

1 **Garden design**

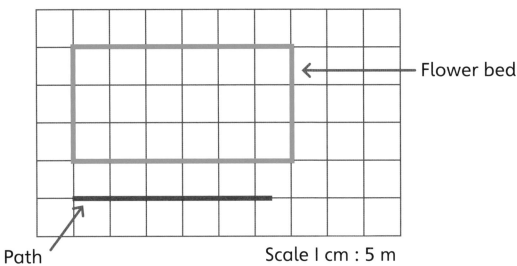

Flower bed

Path

Scale I cm : 5 m

a) What is the length and width of the flower bed in real life?

Use the scale to help you.

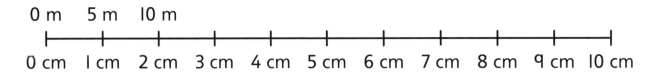

0 m   5 m   10 m

0 cm   I cm   2 cm   3 cm   4 cm   5 cm   6 cm   7 cm   8 cm   9 cm   10 cm

Every ⬚ cm is ⬚ m.

The length of the flower bed in real life is ⬚ m.

The width of the flower bed in real life is ⬚ m.

b) How long is the path in real life?

The length of the path in real life is ⬚ m.

2 Andy is travelling to his grandparents' house.

On the map the distance measures 9 cm.

How far is the distance in real life?

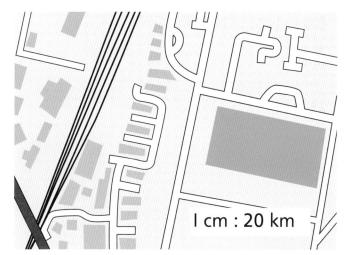

1 cm : 20 km

CHALLENGE

3 Here are two plans of the same house.

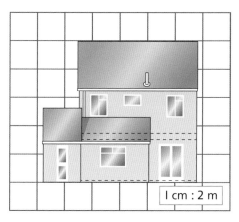

1 cm : 2 m

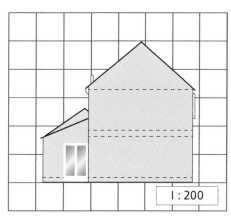

1 : 200

> I think the scales on these houses are different. They are not the same house.

Danny

> I think the scales are the same.

Kate

a) Who do you think is correct?

b) What is the height of the house in real life?

> One of the scales does not have any units. I wonder what that means.

→ **Practice book 6B p165**

Scale factors

Discover

	Height	Shadow length
Kate	130 cm	260 cm
Tree		6·8 m

1 **a)** Kate is 130 cm tall. Her shadow is 260 cm long.

   What is the scale factor?

   b) How tall is the tree?

Share

a) What do you need to multiply 130 by to get to 260?

$130 \times ? = 260$

$130 \times 2 = 260$

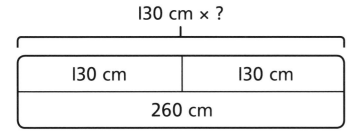

130 cm × ?

130 cm	130 cm
260 cm	

I found the difference between Kate's shadow and Kate. The shadow is 130 cm longer than Kate. Is 130 the scale factor?

Kate's shadow is two times longer than her.

The scale factor is 2.

Scale factor means how many *times* bigger the shape is. I think you need to multiply.

b) The tree's shadow is two times longer than the tree.

Height of tree × 2 = height of shadow

$? \times 2 = 6·8$ m

$6·8$ m $\div 2 = 3·4$ m

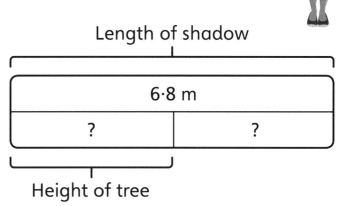

Length of shadow

6·8 m	
?	?

Height of tree

The tree is 3·4 m tall.

Think together

1 Class 5 measure the length of the shadow of a pole at different times of the day.

The pole is 60 cm tall.

a) What is the scale factor when the shadow measures 120 cm?

$60 \times \boxed{} = 120$

$120 \div 60 = \boxed{}$

The scale factor is $\boxed{}$.

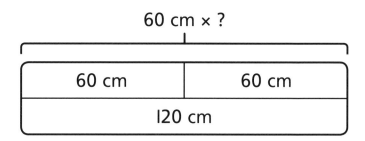

b) What is the scale factor when the shadow measures 180 cm?

$60 \times \boxed{} = 180$

The scale factor is $\boxed{}$.

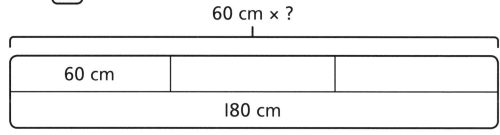

c) What is the scale factor when the shadow measures 420 cm?

The scale factor is $\boxed{}$.

I will draw a bar model to help me work out the answer.

2 A tower is 8 cubes tall.

a) Complete the table to show the height of the tower after these scale factor enlargements.

Scale factor	2	3	4	5	1·5
Number of cubes tall	16				

b) What is the scale factor when the tower is 60 cubes tall?

c) What is the scale factor when the tower is 4 cubes tall?

3

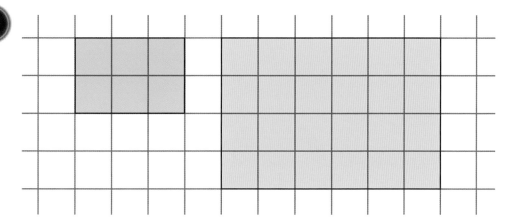

CHALLENGE

The small rectangle has been enlarged by a scale factor of 2.

a) What does it mean to enlarge a shape by a scale factor of 2?

b) What does it mean to enlarge a shape by a scale factor of 3?

c) What does it mean to enlarge a shape by a scale factor of $\frac{1}{2}$?

Draw some examples to explain your answers.

→ **Practice book 6B p168**

Similar shapes

Discover

1 **a)** The two rectangles are similar, but the triangles are not.

What do you think it means for two shapes to be similar?

b) How could you make the two triangles similar?

Share

a) Zac has made two rectangles.

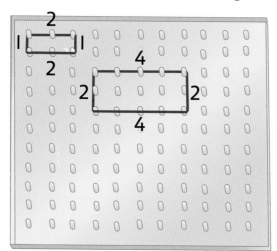

Each side of the larger rectangle is two times bigger than the corresponding side of the smaller rectangle.

So, the rectangles are similar.

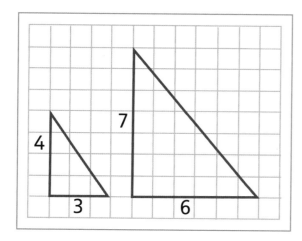

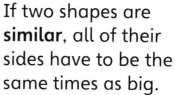

These triangles look similar, as they both have a right angle. I am not sure why they are not similar.

If two shapes are **similar**, all of their sides have to be the same times as big.

$3 \times 2 = 6$

4×2 does not equal 7.

Lexi has drawn two triangles. One of the sides of the big triangle is two times bigger than the corresponding side of the small triangle, but another side is not.

So, Lexi's triangles are not similar.

b) You could make the two triangles similar by making the big triangle 8 units high not 7.

The two triangles are now similar.

Is this the only thing you could have done?

Think together

1 These two rectangles are similar.

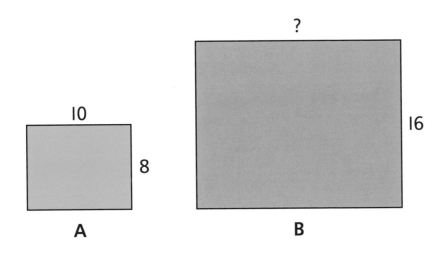

?

10

8

16

A

B

Remember, scale factor means how many times bigger the shape is.

a) What is the scale factor of enlargement?

b) What is the length of rectangle B?

c) What do you think it means to enlarge rectangle A by a scale factor of 3?

2 These two regular hexagons are similar.

What is the scale factor, or enlargement, of a side in hexagon A to a side in hexagon B?

Explain how you know.

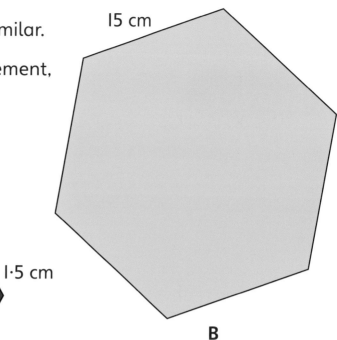

15 cm

1·5 cm

A

B

3 These triangles are similar.

CHALLENGE

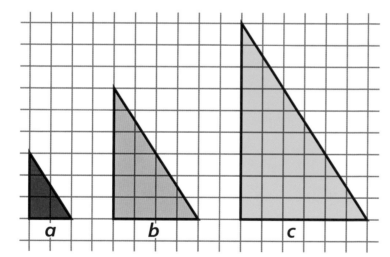

a   b   c

a) What is the ratio of side *a* to side *b*? ☐ : ☐

b) What is the ratio of side *b* to side *c*? ☐ : ☐

c) Describe or draw two shapes with sides that have the ratio 1 : 5.

Do the same for two shapes that are in the ratio 2 : 1.

235

Problem solving – ratio and proportion

Discover

1 **a)** How many grams of curry paste does Toshi need for 6 people?

b) Toshi has 4 peppers.

Does he have enough peppers to make the curry for 6 people?

236

Share

I will find out how much I need for one person and then multiply by 6.

a) Toshi needs 48 g of curry paste for 4 people.

He is making the recipe for 6 people.

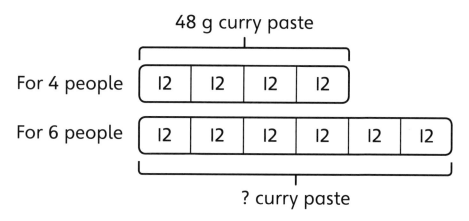

For 1 person, Toshi needs 48 ÷ 4 = 12 g of curry paste.

For 6 people, Toshi needs 6 × 12 = 72 g of curry paste.

Toshi needs 72 g of curry paste for 6 people.

I noticed that what Toshi needs for 6 people is the amount for 4 people, plus half.

b) The recipe uses 3 peppers for 4 people.

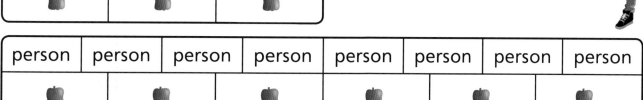

$\frac{1}{2}$ of 3 peppers is 1·5 peppers.

3 + 1·5 = 4·5 peppers

Toshi needs 4·5 (4 and a half) peppers. He does not have enough to make the curry for 6 people.

Think together

1 Olivia is following this recipe.

Makes 12 cupcakes

Vanilla Cupcakes

240 g flour
3 eggs
300 g butter
120 g sugar

a) How much flour does she need to make 15 cupcakes?

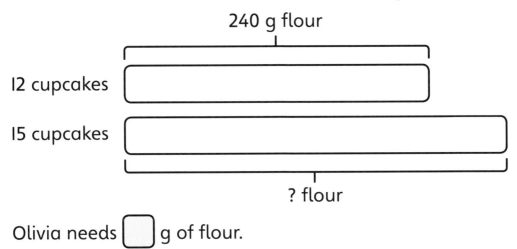

240 g flour

12 cupcakes

15 cupcakes

? flour

Olivia needs ⬚ g of flour.

b) How much butter should Olivia use to make 10 cupcakes?

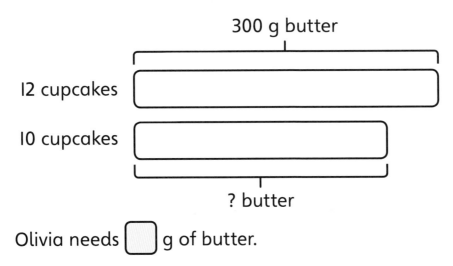

300 g butter

12 cupcakes

10 cupcakes

? butter

Olivia needs ⬚ g of butter.

238

c) If Olivia has 200 g of sugar, how many cupcakes could she make?

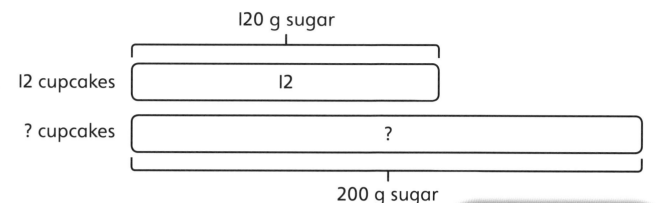

120 g sugar

12 cupcakes | 12

? cupcakes | ?

200 g sugar

Olivia could make ☐ cupcakes.

I wonder if I always need to work out the cost of 1 ticket to work some of these out.

2 It costs £7·50 for 3 people to visit an art gallery.

Fill in the missing numbers to complete the table.

Number of people	3	6	15	30	60
Total cost					

3 A club has booked a tennis court for a set amount of time.

There are 6 pairs of players that need to play.

This means each pair gets 20 minutes on the court.

a) If there were 3 pairs, how long would they each get on the court?

b) If there were 8 pairs, how long would they each get on the court?

CHALLENGE

I think the answer to part a) is 10 minutes because 3 is half of 6 and 10 is half of 20.

I think you need to think differently for this question. I am going to work out the total time they have booked the court for.

239

Problem solving – ratio and proportion ➋

Discover

I **a)** How much water does Sofia need to add to 350 ml of tomato feed?

b) Sofia gives the large tomato plant twice as much feed as the small plant.

If she gives the two plants 1,200 ml of feed in total, how much feed does each plant get?

Share

I wrote it as a ratio and simplified by dividing.

a) 200 ml of tomato feed is needed for every 800 ml of water.

Method 1

This means that for every 1 ml of plant feed, Sofia needs 4 ml of water.

To work out how much water Sofia must add to 350 ml of feed, we multiply by 350.

Sofia needs to add 1,400 ml of water.

Water : Feed

÷ 200 ⟲ 800 : 200 ⟳ ÷ 200
4 : 1

Water : Feed

× 350 ⟲ 4 : 1 ⟳ × 350
1,400 : 350

Method 2

800 ml water for 200 ml of plant feed.

400 ml water for 100 ml of plant feed.

200 ml water for 50 ml of plant feed.

Water

100 ml	100 ml	100 ml	100 ml	100 ml	100 ml	100 ml	100 ml
100 ml				100 ml			

Feed

Water

100 ml	100 ml	100 ml	100 ml	100 ml	100 ml	100 ml	100 ml	100 ml	100 ml	100 ml	100 ml	100 ml	100 ml	100 ml	100 ml
100 ml				100 ml				100 ml				50 ml		50 ml	

Feed

So, for 350 ml of plant feed we add these up:

800 + 400 + 200 = 1,400 ml of water

Sofia needs to add 1,400 ml of water to 350 ml of tomato feed.

I drew a bar model to show the comparison and scaled it up.

b)

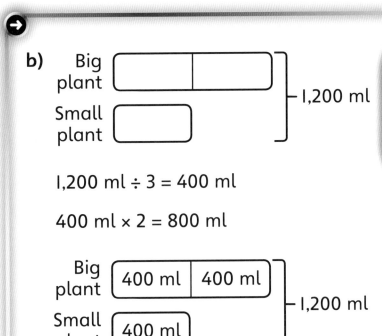

I drew a bar model where the bar for the big plant was twice as long as the bar for the small plant. I know in total Sofia uses 1,200 ml of feed.

1,200 ml ÷ 3 = 400 ml

400 ml × 2 = 800 ml

The big plant gets 800 ml of feed.

400 ml × 1 = 400 ml

The small plant gets 400 ml of feed.

Think together

1 At a holiday club, the ratio of children is 2 boys for every 3 girls.

There are 27 girls at the club. How many boys are there?

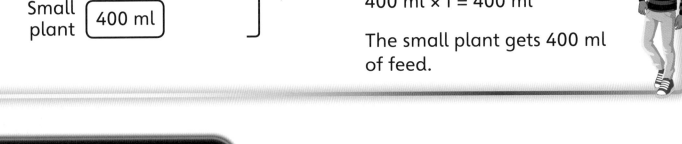

Choose your method or try both. Which one do you prefer? Can you see the connection?

Boys : Girls

2 : 3

? : 27

There are ☐ boys at the club.

2 Amelia has two parcels to post.

One weighs three times as much as the other.

If the lighter parcel weighs 110 g, how much do the parcels weigh in total?

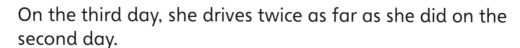

Lighter parcel — 110 g

Heavier parcel

?

The total weight of the parcels is ☐ g.

3 It takes a lorry driver 3 days to travel 784 km.

She drives twice as far on the second day as she did on the first day.

On the third day, she drives twice as far as she did on the second day.

How far does she drive on the second day?

CHALLENGE

I will draw a bar model with three bars.

I wonder how many equal parts there will be in your bar model once you have drawn it.

→ **Practice book 6B p177**

End of unit check

1 On a farm, for every 1 cow there are 4 sheep.

There are 24 cows. How many more sheep than cows are there?

A 4   B 72   C 24   D 96

2 Max measures his goldfish. It is 12·5 cm long.

It is five times bigger than when he bought it.

How big was the goldfish when Max bought it?

A 2·5 cm   B 7·5 cm   C 17·5 cm   D 62·5 cm

3 In a school, the ratio of children having a school dinner to children having a packed lunch is 2 : 3.

There are 240 children in the school. How many have a packed lunch?

A 48   B 96   C 144   D 160

4 On a map scale, 1 cm represents 2 m.

Which of these statements is not true?

A A distance of 5 cm on the map is equal to 10 m in real life.

B A distance of 1 cm on the map is equal to 200 cm in real life.

C This is the same as the scale 1 : 200.

D A distance of 50 m in real life is equal to 100 cm on the map.

5 Shape A is enlarged to create shape B.

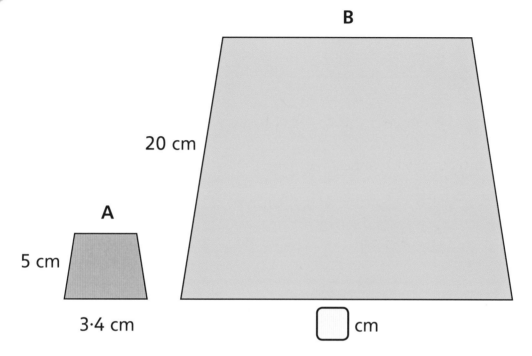

What is the missing measurement?

A 5·5 cm   **B** 6·8 cm   **C** 13·6 cm   **D** 34 cm

6 8 identical cubes and 5 identical spheres have a total mass of 200 g.

The mass of 4 cubes is 80 g.

What is the mass of 7 cubes and 2 spheres?

7 In a pattern, for every 7 circles there are 3 squares and 2 diamonds.

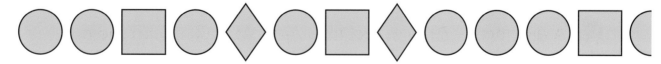

There are 63 circles in the pattern.

How many shapes are there altogether?

→ Practice book 6B p180

Wow, we have solved some difficult problems!

Yes, we have! Can we find even better ways to solve problems?

What have we learnt?

Can you do all these things?

⚡ Multiply and divide decimals

⚡ Find percentages

⚡ Solve equations

⚡ Convert metric measures

⚡ Find the area of triangles and parallelograms

Some of it was difficult, but we did not give up!

Now you are ready for the next books!

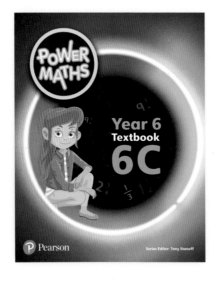